To Rosalie

# Contents

# Preface

For the past fifteen years I have been a welding engineer, concerned with the invention, design, and development of new processes and equipment. Prior to this I spent some thirty years as a practical welder, working on such diverse projects as railway rolling-stock fabrication, pressure-vessel construction, armoured vehicles and aircraft, as well as motor-vehicle manufacture.

I therefore have something of a sense of dedication in writing this book. I want to share, within the limitations of the format of this book, my experiences in welding with you, the student.

The real secret of success is *enthusiasm*

If this book helps to instil this, then its purpose will have been accomplished — as will yours, for you will be reaching out to success . . .

I would like to express my sincere appreciation to B.O.C. Ltd, E.S.A.B. Ltd, A.G.A., U.K. Ltd, the Lincoln Electric Co. Ltd, C. S. Milne and Co. Ltd, and James Neill and Co. Ltd, for the use of photographs appearing in this book. My sincere thanks and appreciation also go to Mrs Valerie Bryon, who typed the manuscript, and to Ruby, for her inspiration.

Peter F. Woods

# 1. INTRODUCTION

## GENERAL SAFETY

Constant reference to safety is made throughout this book. Certain aspects are amplified in appropriate sections (for example, acetylene and oxygen are dealt with specifically in part 3). Handling equipment should never encourage anything but respect, and careless or unauthorised use of hand or machine tools may not only endanger your own safety but that of the people around you.

*It is dangerous to:* interfere with machinery with which you are not familiar; use tools or equipment for unauthorised purposes; conduct experiments with apparatus or materials without being instructed to do so.

*Some of the main causes of workshop accidents are: carelessness, ignorance, risk-taking* or interference with apparatus or machinery; *wrong use of materials,* for example, using hardened tools, like files or drills, as hammers. *Untidyness,* for example, oil, grease, nuts, bolts, or metal parts on the floor could cause people to stumble and fall against sharp edges, machinery, or glass. *Using any gas or combination of gases, including air, for any purpose other than that recommended is dangerous;* for example, the impact of a high-pressure air jet on the skin can produce an embolism and death; carbon dioxide, nitrogen and argon can cause death by suffocation in unventilated spaces.

Oxygen, acetylene, propane, hydrogen, natural gas, coal gas, carbon monoxide and liquid hydrocarbons like petrol, volatile oils and paraffins are all explosion and fire risks if handled wrongly.

## Cutting Edges

These can cause severe injury and the first rule is always to keep both hands behind the cutting edge. Where machinery is concerned, for example, saws, guillotines, lathes, shapers, millers or grinders, a further point should be noted

*Never operate any machine without making sure that the safety guard is correctly fitted and in good order*

## Injuries through Flying Objects

Into this category fall damage to the person caused by molten metal — for example, from a spot-welding machine or from flamecutting — splinters, swarf, or grinding sparks. *Wear the recommended safety clothing.* Safety clothing includes goggles, aprons, boots, gloves, welding helmets, leather caps and approved safety footwear.

## Correct Dress

Items of dress such as ties, belts, unfastened shirt-sleeves or coats can be the cause of severe injury or fatal accidents. For example, a dangling tie or shirt sleeve could wrap around a rotating shaft and pull you into machinery. Long hair is another hazard; caps should always be worn; remember too that long hair is a fire risk when you are engaged in welding.

## Injuries through Ignorance or Abuse

Do not take risks with anything with which you are unfamiliar — and even if you are familiar with it, *do not take risks. Ask* if you are unsure; the person to ask is your instructor — he has the right answers.

Wilful misuse of equipment in a misguided spirit of fun may cause severe injury or death to you or the people you work with. Cylinders of any compressed gas or liquid can become bombs if wrongly treated; almost any material can be lethal if used incorrectly: electricity can kill, water can drown, gases can poison or suffocate, a single match can burn down a factory. *Think,* and if in doubt, *ask.*

Many apparently harmless materials can be highly dangerous under certain conditions. For example, rubber, flour, sugar, metal and cork (to name but a few common substances) can produce violent explosions if mixed with air or oxygen. Some rare-earth metals will spontaneously ignite if sprinkled in the air. Vapour or powder residues in containers (oil drums are a classic example) are capable of producing explosions if ignited. Even a soldering iron can initiate an explosion with volatile liquids like high-octane petrol.

To apply a heat source of any kind to a vessel is potentially dangerous, no matter whether the vessel is full or empty. Filling an exhausted container, tank or vessel with water, leaving only the immediate working area out of contact, is usually safe. Large vessels require steam-cleaning which may in some cases take as long as twenty-four hours, depending on the nature and amount of contaminating substance to be removed. Even after this, it is best to ventilate the vessel with air for a few hours.

## Electrical Safety

Even low voltage can kill under certain circumstances, so *never* believe it is safe to take risks with electricity. Never interfere with any electrical apparatus; if something has gone wrong, report it to your supervisor.

## Electric-shock Treatment

*To free the victim from contact with the electricity*, switch off the current immediately or get someone to do so; do not attempt to remove a person from contact with a high-voltage source unless suitable equipment, insulated for the system voltage, is available. When attempting to free a person from contact with low or medium voltage use rubber gloves, rubber boots and a rubber mat, or an insulated stick, but if these are not available use a loop of rope, a cap, or a coat to drag the person free. Whatever is used must be non-conducting.

*After release*, do not waste time moving him. Lay the patient down on something dry, if possible, and if there is no sign of breathing *immediately* proceed with artificial respiration and send someone for a doctor and ambulance.

*Artificial Respiration: Mouth-to-mouth Method*
Remove any foreign material — false teeth, vomit, etc. — that may be blocking the air passage. To open the air passage, tilt the patient's head back as far as possible. Use one hand to push the patient's head backwards and the other to pull the jaw forwards, thus slightly opening the patient's mouth.

Take a deep breath, place your mouth over the patient's mouth and blow. Then press firmly on his chest to expel the air. Repeat this about five times a minute.

*External Cardiac Massage*
If a patient's condition *does not improve* after one full minute of apparently adequate artificial respiration and you can see that the patient's lips remain blue and his skin pale and you feel sure that the heart has stopped beating effectively,

external cardiac massage may be necessary. If a doctor or nurse is present he or she will perform this. If someone who knows how to perform external cardiac massage is in attendance, this task should be delegated to him. For further details see the electricity regulations contained in the Factories Act, 1961.

## Lifting or Moving Materials

Injuries to the spine can easily be caused by incorrect body-posture when lifting. The correct stance is

    1.  Feet slightly apart
    2.  Knees slightly bent (so that the weight will be taken by the powerful leg and thigh muscles)
    3.  Take hold of the load, keeping the back straight
    4.  Lift by straightening the legs

## Fire Hazards

Before welding or cutting in any environment obtain authorisation from your supervisor. Note where fire extinguishers are kept; read operating instructions; make sure that fire emergency exits are unobstructed; warn people in the area before starting work; use the correct dress and equipment.

## PRINCIPLES OF FUSION

The joining and fabrication of metals by brazing and casting are ancient arts whose origins are lost in remote antiquity — yet they are still used today, although in more diverse and sophisticated ways. The principles of *forge welding* for example, are utilised in resistance spot-welding but with electrical-resistance heating taking the place of the blacksmith's hearth and an air cylinder doing the work of his hammer. (For a detailed description see the section on resistance welding, p. 98.) In forge welding, metal is heated to a temperature at which it becomes soft and malleable, and the parts are then welded together by hammering. (Forge welding is not the same as *forging*, which is a method of shaping parts in a die or on an anvil using power- or hand-operated hammers or rams.)

For centuries the only available source of heat for metal working was provided by the burning of charcoal or wood and not until much later by coal or coke. The temperature of the early blacksmith's forge was increased by forcing air through the fire with hand- or foot-operated bellows, this task being undertaken usually by the apprentice.

The first inkling of things to come occurred when oxygen was isolated by Joseph Priestley in 1774, who prepared it by heating mercuric oxide. By the mid-nineteenth century, progress was being made in steel making and the possible use of the electric arc for welding. At the turn of the century, the Brin brothers tried to produce oxygen on a commercial scale by heating barium oxide but the process was abandoned in favour of the liquifaction and fractional distillation of air, a process in which control of temperature is used to separate the component gases of air. This method was first used on a large scale by C. Linde in 1902, in Germany.

The gas was mainly used for medical purposes and as a source of illumination in conjunction with lime, to give 'lime-light'. Then came a more significant discovery — it was found that by mixing oxygen with coal gas the flame temperature produced enabled iron to be melted and cut with ease. Unfortunately, the first people to take advantage of it in the United Kingdom were the criminal fraternity in Warrington where they used it to open a safe. The event caused a sensation and the authorities were quite alarmed at the threat that this new 'secret weapon' posed against security.

Acetylene was first discovered by Edmund Davy in 1836. In 1892 T. L. Willson, a Canadian working in the United States, while trying to smelt aluminium ore using a mixture of bauxite, coke and lime, discovered a technical process of acetylene production. (His experiment in producing aluminium by electrolysis was a failure and he threw the material away, but some days later he became aware that the substance was emitting a gas which, on subsequent analysis, proved to be acetylene, $C_2H_2$.) This first found use in lighting and was manufactured by adding water to calcium carbide in a small vessel or gasometer. Eventually a blowpipe was developed for the controlled combustion of oxygen and acetylene and a new era in the fabrication of metals had arrived. Early attempts to compress acetylene met with disastrous results until 1897 when the chemist G. Claude discovered that the gas was soluble in acetone and the solution could safely be absorbed in a porous material like asbestos, carbon, or diatomaceous earth, packed into a cylinder. About the same time, the fractional-distillation technique for manufacturing and storing oxygen as a liquid was developed. Fractional distillation of liquid air is still the method used today to obtain oxygen, as well the gases argon, krypton and neon.

Arc welding was first investigated in 1881 by Auguste de Meritens who experimented with arcs using carbon rods, and in 1888 the Russian, N. G. Slavianoff, developed a process using a consumable steel electrode. This technique was employed for many years, its main applications being the repair of blowholes, etc., in steel castings. Arc welding really came into prominence commercially with the invention (by Oscar Kjellberg of Sweden) of the flux-coated electrode in 1907. Arc welding was, up to the Second World War, something of a Cinderella in engineering but the war forced its use and development by sheer necessity. Today a whole family of processes have grown out of the original concept of metal-arc welding and much more development lies ahead.

## Heat and Work

In the use and understanding of every branch of sciences and arts, there are disciplines and laws that must be observed; fusion welding is no exception. It is therefore most important that we have a fundamental appreciation of thermodynamics, perhaps better known as the science of heat and work.

Let us begin by describing a simple experiment that illustrates the principle of heat flow or *thermal conduction*. Figure 1.1 shows a tank containing hot water; projecting from the sides of the tank are metal bars and on the end of each one is a piece of paraffin wax. Heat flows along the bars, at a rate depending on the thermal conductivity of the particular metal, eventually melting the wax. Next to silver, copper is the best conductor and compared with these, iron and lead, for example, are poor conductors.

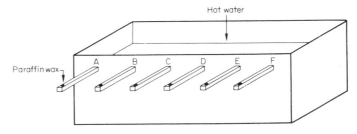

*Figure 1.1    Thermal conduction in solids; A — copper, B — zinc, C — brass, D — lead, E — slate, F — wood. Thermocouples connected to galvanometers may be used instead of wax*

Heat is not necessarily transmitted by conduction alone — it can also travel by *convection* and *radiation*. Heat is conducted at a constant rate through a homogeneous solid; the rate of heat transfer, mostly by convection, in a liquid or a gas is not constant.

Figure 1.2 shows hot liquid rising and the colder liquid sinking because of its greater density. In practice it is by no

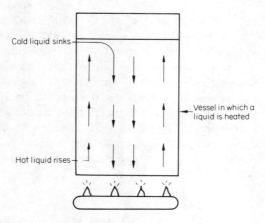

*Figure 1.2    Heat transfer in liquids*

means as simple as the illustration seems to suggest — this only outlines the basic principle.

Heat always travels from a hot body to a cooler one, and in the process it may be made to perform work, for example, in the steam engine. A refrigerator, on the other hand, extracts heat from a hot body by means of work put into the system (by the pump), but the basic principle is still the same.

The first and second laws of thermodynamics may be stated as foilows.

1. When any closed system is operated through a cycle (that is, a sequence of events in time), the net work delivered to the surroundings is proportional to the net heat taken from the surroundings.

This is the law of the conservation of energy, which means that if you put a measured quantity of energy into a system you can only get the same amount (theoretically) out of it. This is the law that would-be inventors of 'perpetual-motion' machines always fail to take into account!

2. It is impossible to construct a system that will operate a cycle, extract heat from a reservoir and do equivalent work on the surroundings.

Broadly interpreted this means that it would not be possible to propel a ship using only the heat of sea water or generate electrical energy in a power station by absorbing the heat of the atmosphere. This is because (remembering what we said about heat always flowing from a hot body to a cooler one) there is no large cold body available, relative to the sea or the atmosphere; thus the thermal differential is insufficient to provide a significant source of heat or work.

## The Thermal Reservoir

Solids, liquids or gases can be regarded as thermal reservoirs in which heat is constantly ebbing and flowing. Metals (with the exception of mercury) are, under normal conditions of temperature and pressure, frozen liquids.

Heat travels through solids according to the characteristic thermal-conductivity of the material as shown in figure 1.1. Heat reaching the surface of a solid flows away from it at a rate determined by the thermal conductivity of the surrounding medium which may be gas, liquid, solid or a combination of them. Heat may be dissipated by conduction, radiation or convection. If copper and iron blocks are heated to the same temperature, the rate of heat loss is regulated by the rate at which heat travels through the blocks and the surroundings. Because heat travels faster through the copper, the rate of heat input required to maintain a constant temperature will be greater than for iron. Figure 1.3 shows the principle of heat flow, the blocks on the left are heated to the same temperature, A representing the hot body, and B the cooler one.

1. The rate at which heat travels from A to B is determined by the surrounding medium, gas or liquid.

2. In this case, heat will flow from A to B at a higher rate because the blocks are in physical contract.

3. Because the blocks are joined together by welding, heat will be transferred almost entirely by conduction.

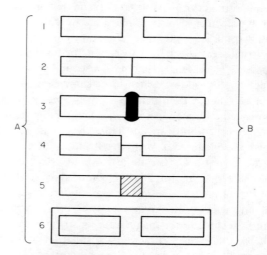

*Figure 1.3    Variable factors in heat transmission; A, hot body, B, cold body, 1 — air gap, 2 — physical contact, 3 — welded joint, 4 — wire bridge, 5 — asbestos bridge, 6 — immersed in fluid*

4. Here, blocks are connected by a piece of wire and although a thermal bridge has been provided much heat will be lost by radiation to the surrounding environment.

5. The two blocks are separated by asbestos which is such a poor thermal conductor that it is often used as an insulator.

6. Since both blocks are immersed in a fluid, the rate of heat transfer between them will depend on the thermal conductivity of the liquid; thus if the blocks were immersed in molten lead then the rate of heat transfer between them would be much greater than if they were surrounded by wax or oil.

## Expansion and Contraction

Metals generally expand when heated but there are some exceptions to the rule, such as printers' metal which expands as it cools. Cooling metal normally contracts, giving rise to the often undesirable phenomenon known as distortion. Iron heated below its melting point becomes plastic and can be permanently deformed. Some materials, like cast iron or glass, react to heat-induced stresses by cracking unless great care is taken. When heating these, it is essential to build up heat slowly and uniformly and to take just as much care during the cooling-down period.

Care must be taken to heat metal parts without causing excessive distortion and a uniform input should be provided when possible. The forces generated by expansion or contraction are very powerful. For example, the expansion joint in a railway line is so called because it is designed to compensate for the increase in length when the rail is strongly heated by the sun. If no gap were provided, the rails would buckle. When uniformly heated, metal expands equally in three dimensions as shown in figure 1.4. The sphere is made to close dimensional tolerance so that it will only just pass through the ring at room temperature.

Metal cooling in a mould, as in the case when producing sand castings, shrinks to form a depression at the point of pouring (see figure 1.5) so provision is usually made for topping up the casting from a ladle.

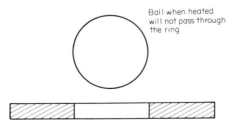

Figure 1.4    Thermal expansion

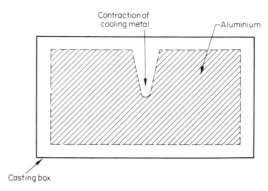

*Figure 1.5    Shrinkage of molten metal*

## IRON AND STEEL

The metal iron (symbol Fe, from the Latin *ferrum*) is, in the pure form, ductile, soft and malleable, having a melting point of 1535 °C. Cold sulphuric and hydrochloric acids do not attack it readily in this state. The easiest method of preparing pure iron is by reduction of the precipitated ferric oxide in an atmosphere of hydrogen. The metal has an atomic weight of approximately 56 and a specific gravity of 7.86.

At red heat, iron oxidises rapidly, producing a layer of oxide on the surface. Red-hot iron will decompose water to liberate hydrogen according to the reaction

$$3Fe + 4H_2O = Fe_3O_4 + 4H_2$$

The principal sources of iron are its ores: magnetite, $Fe_3O_4$; red haematite, $Fe_2O_3$; specular iron ore — a crystalline variety of red haematite; brown haematite — a hydrated ferric oxide; siderite (or spathic iron), $FeCO_3$; clay ironstone; blackband ore; and iron pyrites, $FeS_2$, commonly known as 'fools' gold' because of its resemblance to the precious metal.

In the making and refining of iron from its ores, the process chosen largely determines the quality of the product. In the making of cast iron, pig iron and scrap are used in the cupola furnace which contains a mixture of iron and coke as fuel with limestone as a fluxing agent. The charge is melted by burning the coke in a blast of air, forced through tuyères at the bottom of the vessel. The iron thus produced has a high carbon-content — about 3 per cent; purification and annealing produces, as desired, white, grey or malleable cast-iron.

White cast-iron contains a large amount of iron carbide, known as cementite. The metal has the high compressive-strength characteristic of cast iron but it is hard and brittle. Grey cast-iron is a softer variety containing graphite and having a coarse crystalline structure. Malleable cast-iron is produced by annealling grey cast-iron in a furnace at about

750–800 °C so that the chemically combined carbon becomes dispersed throughout the mass of metal in finely divided form.

Spheroidal graphite iron (S.G.I.), also called nodular cast-iron, is a variety produced by adding small amounts of magnesium to molten grey-iron to produce carbon inclusions in spheroidal form. The metal has high strength and ductility with good casting-potential.

Wrought iron consists of refined metal containing about 2 per cent of furnace slag in the form of stringers in a matrix of iron.

Processes for the making of steel include: the acid process, the basic process, open-hearth process, basic oxygen process, Bessemer converter, induction furnace process, vacuum smelting and oxygen blowing.

The acid process is so called because the refractory furnace-lining of silica does not significantly affect the composition of the metal that is melted, whereas the basic-(lime-)lined furnace alters the metal composition by reaction with its sulphur and phosphorus content.

The open-hearth furnace, which usually has a lining of highly refractory magnesium carbonate, $MgCO_3$, known as magnesite, is capable of producing a variety of steels. Coal or gas may be used as fuel which is burned with pre-heated forced air.

The sequence of operations is: (1) charging; (2) melting; (3) refining; (4) purifying, for which purpose limestone, iron oxide, or iron ore are added as sources of oxygen. Sometimes gaseous oxygen is used to accelerate the transformation.

If oxygen-rich steel is poured into cast-iron moulds the liquid metal starts to solidify close to the inner surface of the vessel in the form of high-purity iron. The carbon in the centre of the cooling metal combines with the excess oxygen to produce carbon monoxide, CO, which rises and burns off at the surface.

Killed steel is produced by adding deoxidising elements such as silicon, aluminium or manganese to the melt. Basic oxygen steel is produced in a converter through which a jet of oxygen is forced. Steel made by this technique takes a far shorter time to produce than with the older processes, but it is only economical for large-volume manufacture because of the higher cost of using oxygen.

The Bessemer converter is not a melting furnace but a vessel lined with refractory material into which molten iron is discharged directly from a furnace. Cold air is then forced through the liquid metal to oxidise impurities. Carbon monoxide thus produced burns at the top of the converter and the flame subsides after about fifteen minutes, signifying the end of the reaction.

Induction-furnace melting is used for the production of high-quality alloy steels in relatively small quantities. The principle is that of electromagnetic induction and is applied by building the melting crucible inside a water-cooled coil of copper piping which is, in effect, the primary of a transformer. High-frequency current passing through the coil induces eddy currents in the crucible charge, eventually melting it.

Vacuum melting is also a process used to manufacture special steels in limited quantity. An induction furnace is used for melting and the unit is enclosed in an airtight vessel from which the air and gases from the melt are extracted.

### Casting and Forge Welding

The casting of iron direct from the furnace into moulds is a centuries-old practice. The usual method is to pour metal from sand-lined ladles into the casting box or mould filled with special sand mixtures. For making castings in iron, bronze or brass with fine detail, the Chinese lost-wax process is still sometimes used.

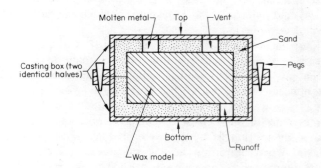

*Figure 1.6    The lost-wax casting*

Figure 1.6 shows, in simple form, the method used. The wax model is placed in the bottom half of the box; the top is then put on and sand rammed in, with vent holes provided through which the metal will be poured and gases escape. The whole assembly is heated until the wax melts and drains out through the runoff hole in the bottom of the box. The lower vent is then sealed off with clay, and metal poured in until it rises to the top of the upper vent-holes. The metal contained in vent holes is surplus to the casting but provides a reservoir that compensates for the reduction in volume of the metal cooling inside the box. When the process is complete, the two halves of the box are dismantled by removing the aligning pegs, and the columns formed in the vent holes are broken off (these spikes of metal are known as 'risers').

Casting was first introduced into England at Buxted, Sussex in about 1543, when cannon and ammunition were produced on a small scale. The process was not widely used until charcoal was replaced by coke, and in 1709 Abraham Darby established the first foundry on a commercial scale at Coalbrookdale in Shropshire. It was here that the famous iron bridge (still existing at Ironbridge, Shropshire) was made, as well as the first steam-engine cylinders.

Malleable iron or low-carbon steel is commonly used for forge welding, and the metallurgical and physical properties of these welds are extremely good. The method is still used to some extent for producing ships' anchor-chains although a process known as 'flash-butt welding' is now more common.

The technique known as *friction welding* (see p. 101) is finding increasing application in the joining of rods, bars and tubular sections, not only in iron and steel but also in a great variety of other metals and alloys.

### BASIC METALLURGY

In welding, fusion or melting is followed by solidification of the weld and the parent metal adjacent to the joint. It is therefore important to know something about the solidification process in pure metal and alloy systems.

In a liquid pure metal the atoms are randomly distributed and vibrate about a mean position. The energy that this vibration requires is supplied by the heat that was originally needed to melt the metal. As the temperature falls these vibrations are reduced in magnitude until a point is reached when freezing begins. Groups of atoms begin to converge at various points, or *nuclei*, in the liquid. These nuclei are ordered arrangements of atoms and consist of three main metal-*lattice* types: body-centred cubic, face-centred cubic and hexagonal close-packed. Figure 1.7 shows the atomic arrangements of the three types. In the body-centred-cubic lattice there is an atom at each corner of a cube and one in the centre of the cube; there is a body-centred-cubic form of iron. In the face-centred-cubic structure atoms are again situated at each corner of a cube with atoms at the centre of each of the cube faces; aluminium and copper freeze with this arrangement. The hexagonal-close-packed lattice is more complex, with atoms at the corners of two adjacent hexagons; examples of this mode of lattice formation are shown by zinc and magnesium. It should be noted that the figures are not drawn completely accurately, in reality all the structures are more compressed, with the atoms touching each other at their peripheries.

As freezing continues these initial nuclei grow and more atoms attach themselves to the original nucleus. The growth is

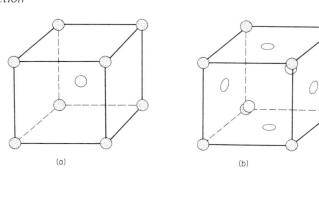

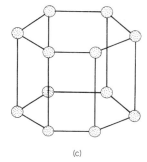

*Figure 1.7   Metal lattice types; (a) body-centered cubic, (b) face-centered cubic, (c) hexagonal close-packed*

in three dimensions but is not in the form of a uniform block of lattices; this is because growth is easier in certain directions owing to the heat-convection currents in the liquid. Thus the structure has a 'spiky' appearance like ice crystals, with arms radiating from the original line of nuclei, giving a *dendritic*, or tree-like formation.

On further freezing the space between the branches is filled in and loses its dendritic appearance. On completion of solidification the different (originally dendritic) blocks will impinge on each other and since they were not formed in the same plane there will be a difference in orientation where two units meet. The last part of the liquid to freeze is in these areas between units and because of the orientation difference the units will share atoms. This sharing is necessary because some of the atoms required to complete the units would otherwise be too close to each other and so a compromise is reached where one atom is used instead of the expected two. Hence the distance between these shared atoms will be greater than that between the atoms in the body of the original solid. When freezing is completed, each unit derived from a dendrite is a *crystal* or *grain* in its own right and the narrow parts between each grain, where the atoms are shared between crystals, are the *grain boundaries*. It can be seen that, compared with the body of the grain, the structure at each grain boundary is imperfect due to the differing atomic distances and this causes the grain boundaries to exhibit very special and important properties.

### Alloys

The freezing mechanism described above applies to a single pure metal. For practical applications most pure metals have a very limited range of properties and these properties may be modified by the addition of one or more alloying elements.

During the transformation of fused metal to the solid condition, nucleation (that is, formation of crystals at localised centres in the liquid) and crystal growth occur. The crystalline structure may be dendritic or columnar, depending principally on the cooling rate. The first crystals to separate from the liquid metal may be of a different composition from the mother liquid and will undergo progressive changes as solidification approaches. Therefore while the metal is cooling, the liquid and solid phases may differ considerably in composition. The final composition of weld metal may also be modified by the process of diffusion, but because a deposit usually cools very rapidly, due to the quench effect of the surrounding relatively cold metal, diffusion may be limited. In the case of pre- or post-heating, however, there is more time available and therefore a greater probability of chemical and structural modification in the weld metal.

In steel, the element carbon plays a vital role, and its presence either in compound or element form has a marked effect on the physical properties of the metal. Examination of a prepared specimen of steel under a microscope may reveal segregated dark areas known as pearlite, which is a mixture of ferrite and cementite (the compound $Fe_3C$ of iron and carbon, also known as iron carbide).

There are, of course, other elements that can substantially affect the properties of steel. Among these are the metals manganese, molybdenum, vanadium, chromium and nickel; and non-metallic elements such as silicon, carbon, sulphur and phosphorus, but only in strictly controlled amounts.

As the heating medium (that is, arc or flame) travels along the joint, deposited weld-metal is subjected to rapid thermal cycling, depending on the temperature, thickness and composition of the adjacent metal. The weld metal may be quenched very quickly or relatively slowly, corresponding to a range of conditions from air-cooling to water-quenching.

If the structure being welded is also subjected to a high degree of restraint — for example, because of the inherent inflexibility of its constituent members — the cooling weld-metal will impose contraction forces of high magnitude in the weld area. This stress may be relieved, in ductile metals, by plastic flow which occurs below the melting point, but if high-tensile material is involved the shrinkage forces may cause cracking. This is why pre and post heat-treatments, together with correct welding-conditions must be rigorously applied where indicated, for example, in the repair and fabrication of high-alloy tool and die steels.

The solubility of oxygen and the formation of metallic compounds in weld metal means that protection of the molten pool is necessary. This can be achieved by

1. using a flux
2. providing an inert- or semi-inert-gas shield
3. welding in a vacuum chamber

Oxygen reacts with carbon to form carbon monoxide or with hydrogen to form water vapour. Gaseous oxides trapped in solid weld-metal result in porosity and discontinuities. For this reason the elements aluminium, manganese, silicon, titanium or vanadium may be added to steel because they have a relatively high affinity for oxygen.

Hydrogen is another source of weld-metal porosity and is also a crack promoter. This is why it is so important to make sure that the workpiece is free from contamination and, in the case of metal-arc welding, that the electrodes are dry before use.

### FLUXES

The first rule in welding is cleanliness. Not only must the working environment be clean, for reasons of safety and efficiency, but the workpiece must be clean too.

In welding we are engaged in a constant battle against corrosion, and only the noble metals have a high resistance to atmospheric gases. Engineering metals, like iron, copper, lead, zinc, aluminium and their alloys are constantly attacked by corrosive environments, the most common being air and water vapour.

The rate at which metal is attacked increases with temperature. For example, the formation of the oxides of iron, aluminium and copper is rapid if the metals are strongly heated in air. (When heated in pure oxygen, iron will revert completely to its oxides.) The intensity of reaction between metal and atmospheric gases is greatest when the metal is molten, but below this temperature a progressive combination of metal and oxygen is taking place. A flux or chemical solvent is used to remove the oxide layer. The most common flux used in soft soldering is zinc chloride. However, the situation becomes very complex with alloy materials such as magnesium—aluminium alloys. Aluminium in particular, even at room temperature, forms a very tenacious oxide, with a melting point of about 2200 °C, which is inert to most ordinary chemical solvents.

## Reactive Fluxes

Vigorously acting fluxes based on the halogens, generally fluorides and chlorides, are necessary to deal with the magnesium—aluminium situation. These types of flux are called reactive because they work by producing chemical changes. Most reactive fluxes are hygroscopic (that is, they absorb water) and must be kept in air-tight containers. Because of their acid nature, such flux residues are usually electrically conductive and must be thoroughly removed — in the manufacture of electrical apparatus, damage may be caused by short-circuiting due to trapped flux.

Fluxes generally are corrosive before or after use and it is therefore most important that they are removed before work is assembled. Some fluxes, for example, for brazing, hard soldering, aluminium or cast-iron welding, may be in powder form. Others may be dissolved or suspended in liquids like propyl alcohol, methyl alcohol or aqueous mixtures. Such preparations should be handled carefully. They must not be stored near heat sources, and must be tightly covered when not in use. The fluxes used in metal-arc welding perform their function of oxide removal at high temperature, so they have to exhibit stability under extreme conditions.

Much research has gone into producing effective fluxes and it is most important to use the right flux for a particular job. Fluxes have a temperature—time life. For example, a flux designed for hard soldering (silver soldering) at about 750 °C would rapidly decompose at brazing temperature (950 °C) producing irritant and dangerous fumes.

## Organic Fluxes

Into this category fall resin-based fluxes, stearic acid and chloride-free types. These fluxes are usually not hygroscopic and, compared with the reactive variety, are fairly inert. They are designed to function in lower-temperature ranges and have a shorter temperature—time life. The chief use of organic fluxes is in soft soldering and the low-temperature joining of aluminium and its alloys. All residues must be removed because of the chemical changes that occur after the fluxes are heated to produce corrosive byproducts.

## Fume Removal

In all fusion joining-processes conducted in the atmosphere, fumes are an unwelcome result. Such fumes may include metal particles, oxides and nitrides as well as other constituents, many of which are toxic. Some metals are intensely toxic in combined or finely divided form; these include beryllium, cadmium, nickel and their compounds. The effective ventilation of the working environment is therefore of prime importance.

In addition to fumes associated with metals and their compounds, such materials as silica, asbestos, cellulose, glass, plastics and rubber are hazardous in finely divided form. The byproducts of the welding process may be toxic gases like carbon monoxide, the oxides of nitrogen, ozone, acid or alkali vapours, etc. It is important that extraction equipment provided in a workshop is fitted with filters for removing contaminating material before the air is returned to the area.

Vacuum extraction is, to date, the most successful method of air cleaning; the extraction equipment may be sited locally or generally. Each application has to be carefully studied, thus for large fabrications employing several welders simultaneously, general extraction would be most suitable; for one operator working in a booth and producing small repetitive assemblies, local extraction would suffice.

Great care is necessary when working in confined spaces or vessels. In these circumstances, there is not only the hazard from fumes but the less apparent ones associated with the welding process: carbon dioxide for example, can quickly suffocate; oxygen can cause a rapid and often disastrous fire; acetylene, propane, natural gas, butane and other inflammable gases can cause fire and explosion.

*Always check welding equipment for gas leaks before entering a confined space or vessel*

Before entering a vessel make sure it has been certified as being safe. If it is necessary to enter through a manhole and this is the only exit, wear a lifeline and make sure someone is standing outside ready to pull you out in an emergency. An airline should be used to extract fumes generated by welding or cutting if portable vacuum-extraction equipment is not available. *Never* use an airpipe to cool yourself — compressed air can penetrate the skin causing an embolism and death.

## ENGINEERING MATERIALS

Two main classifications of material can be made — *metallic* and *non-metallic*. The largest metallic group is that based on iron or containing a significant proportion of the metal in alloy form. There is a third important group called *cermets*, that are mixtures of metals and ceramic materials.

### Sintered Materials

Metals in finely divided form (usually achieved by electrolytic or chemical processing) are intimately mixed, compressed into dies and heated to form shapes. The *sintering* is done in an electric furnace using an inert atmosphere.

Iron is rarely used without the addition of other elements, such as chromium, nickel, molybdenum, manganese, carbon (a non-metal), vanadium, etc. The addition of these confer special properties such as hardness, high tensile- or compressive-strength and resistance to corrosion.

### Non-ferrous Metals

Among these, such metals as aluminium, zinc, nickel, copper and titanium can be used in the unalloyed condition. Aluminium, for example, is often used for domestic utensils, cooking — as foil — and in wire form. Zinc is commonly used for galvanising and corrosion-resistance treatment; titanium too, because of its inertness to certain chemicals, is increasingly used in chemical plant.

### Ferrous Alloys

There is an enormous range of combinations of metals (that is, alloys) possible to obtain specific properties, but those that contain a proportion of iron, or are based on iron form the largest group in general engineering. The iron—carbon alloys are probably the best known among them. *Steel* is the name given to this group, which is further defined as being a range of iron—carbon alloys containing about 0.1 to 1.7 per cent carbon.

The quantity and state of the carbon in iron determines the properties of a specific alloy. For example, iron containing about 0.005 per cent carbon will exist as a solid solution at room temperature (ferrite) the remaining carbon forming iron carbide or cementite. Ferrite and cementite form a lamellar structure called pearlite. In an iron—carbon alloy heated to a high temperature a rearrangement of the atomic structure takes place and the amount of carbon in solution increases. This carbon-rich solution (up to 1.7 per cent carbon in iron) is called austenite. The change in the iron lattice is reversible, depending on whether the temperature is increased or decreased.

### Non-metallic Materials

Glass, synthetic resins and ceramics are typical non-metallic materials. Unlike metals, they are poor thermal and electrical conductors but are relatively good insulators. Among them the ceramics are outstanding insulators, being used in high-voltage electrical systems, jet-engine components and welding applications (for example, backing bars). Other important properties of ceramics are their high hardness and resistance to chemical attack over a wide range of temperature. Asbestos, a naturally occurring mineral, also has good thermal- and electrical-insulation properties; because it is fibrous it can be woven to make such items as gloves, capes, aprons and sheets; it can also be compacted to produce a wide variety of goods like vehicle brake-shoe linings and builders' hardware. The synthetic resins, of which there are a great number each with specific properties, have an equal variety of applications. Among these materials are acrylics (used for dental work), polycarbonates (used for general engineering applications), and the epoxies — some of which constitute modern high-strength adhesives.

## HEAT TREATMENT

Properties such as hardness, toughness, crack resistance, etc., are achieved by heating and controlled cooling of carbon steel. Water, oil or air are the common means of controlling the quench or cooling rate. Furnace, salt-bath or induction-heating methods are used for large-scale heat-treatment.

### Annealing

Cold-working, that is, hammering, bending, press forming, etc., causes deformation and stress concentration. Welding promotes further stress concentration, and in order to relieve it, annealing is used. In this process, the workpiece is heated under controlled conditions to a point above the recrystallisation temperature and then cooled slowly. Heat-treatment procedure must be related to the carbon content of the steel.

### Normalising

To machine or temper some steels the temperature of the steel is raised to produce a wholly austenitic structure and then air-cooled. The amount of carbon present determines the temperature and the cooling cycle is calculated according to material thickness. Slow cooling (normalising) of austenite produces ferrite, but rapid cooling, for example, quenching, will transform the austenite into fine needles of martensite, the presence of which makes the steel hard and brittle.

## Tempering

After hardening, carbon steel reheated to around 650 °C (depending on carbon content) will show increased ductility and toughness but reduced hardness, the martensitic structure being transformed into troostite or sorbite, depending on temperature and cooling time.

## Surface Hardening and Case Hardening

A steel—carbon alloy containing, say, no more than 0.15 per cent carbon will not produce a significant increase in hardness when heated. Carbon-rich compounds are used to increase the carbon content of the material surface-area by heating to about 900 °C. The thickness of the case formed depends on the time of exposure to the carbon-rich environment which may be gas, solid or liquid. Sodium cyanide is often used in salt-bath form for case hardening.

## Nitriding

Nitrogen in gaseous or compound form is also used for hardening steels. In this case, compounds known as nitrides are formed. In order to respond to nitriding, however, the material should contain a small percentage of aluminium or chromium.

## SURFACE PROTECTION OF METALS

Synthetic resins, ceramics, paints and rubber are among the non-metals used for surface treatment, the choice depending on the service requirements of the article. Zinc, cadmium, lead, tin, chromium and copper are commonly used for surface coatings. The methods of application include metal spraying, electroplating, dipping, diffusion and chemical deposition. Metallic compounds are also used; lead, zinc and chromium oxides, for instance, can be applied in paint form.

## Cladding

A thin sheet of protective metal can be bonded to both sides of the plate to be protected, by rolling or by explosion techniques. These techniques are relatively expensive and therefore restricted to such applications as vessel lining and protection from corrosion.

## Chemical Coatings

Bonderising and Parkerising are proprietary applications of the acid phosphating-process which entails the heating of the workpiece in a controlled solution of orthophosphoric acid. Small components such as nuts, bolts, screws and washers are often protected in this way.

## Anodising

Aluminium rapidly forms its own protective coating of oxide when exposed, even at normal temperatures, to air. In the anodising process the oxide-layer formation is accelerated by electrolytic action and dyes (which are readily absorbed by the oxide coating) may be added to give a range of attractive coloured finishes.

## Encapsulation

The complete immersion or encapsulation of components in a synthetic resin is a modern technique increasingly used for metal protection. Electronic circuitry is particularly suitable for this treatment.

## Metal Spraying

This process has many applications besides that of protecting metal surfaces but these are too numerous to catalogue here. Metal spraying, using oxy-acetylene as a heat source (with air as a metal propellant and atomiser), can be used for spraying zinc, aluminium, brass, lead, copper, stainless steel and many other metals and alloys. The surface bond is mechanical, the sprayed metal adhering because of keying that has been produced by shot blasting the work.

## Powder-metal Spraying

This is a variation of the oxy-acetylene technique in which a specially prepared, finely divided metal powder is used instead of wire. With this process, fusion of the sprayed metal to the workpiece is achieved.

## Plasma-arc Spraying

The intense local heating produced by the (restricted) plasma arc is ideal for producing a metallurgically bonded protective coating on metals. High-melting-point ceramics can also be sprayed using this technique.

## ELEMENTS OF DRAWING

Before they have learnt the meaning of letters and words, young children learn to understand things by recognition of sounds and images. If you wanted to describe something that you had seen to someone who could not speak English, how could you do this? You could show them a photograph or you could draw it. If you drew the outline of a fish, for example, your basic meaning would instantly be transmitted to them.

One method of giving more information is by the use of *isometric* drawing — see figure 1.8a — this conveys more in-

formation than the *plan*. (Reference should be made to BS 308:1972 Engineering drawing practice.) It does not, however, indicate the shape and depth of the centre hole. To do this we have to use the plan, *front* and *side elevations*. This system of illustration is called *first-angle projection*. Another way in which this can be shown is by using *third-angle projection*. The two methods are compared in figure 1.8b. The pictorial view on the right shows the general contours of the block, the dotted lines representing the edges not visible from the angle of observation.

Before proceeding from this point it will help in further understanding if we consider some of the basic geometric requirements upon which *orthographic projection* (*line drawing*) is based.

**Geometrical Definitions**

A straight line is the shortest distance between two points. Two straight lines intersecting each other to form four equal angles produce four right angles.

A plane figure bounded by four straight lines is a *quadrilateral*. If one pair of opposite sides are parallel the

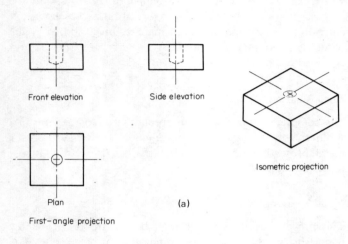

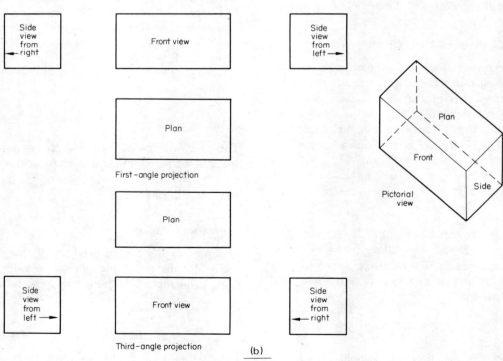

Figure 1.8    *Isometric, first-, and third-angle projection*

figure is a *trapezium*. If a quadrilateral has both pairs of opposite sides equal and parallel it is called a *parallelogram*. If a figure has all its angles as right angles it is a *rectangle* and if all sides of the rectangle are of equal length, it is a *square*. An area bounded by more than four sides is a *polygon*. A plane figure defined by three straight lines is a *triangle* and if two of these sides are equal it is an *isoceles triangle*. If all three sides are equal it is an *equilateral triangle*. If the sides of the triangle are unequal it is called a *scalene triangle*. A triangle that is right-angled has the side of longest length opposite the right angle; this side is known as the *hypotenuse*.

A plane curve whose distance (the *radius*) from a fixed point (the *centre*) is constant is a *circle*. Irrespective of the size of a circle, the *diameter* (twice the radius) has a fixed ratio to its *circumference* and this ratio designated by the Greek letter $\pi$ (pi). For most calculations the value of $\pi$ may be taken to be 3.142. Any portion of the circumference of a circle is called an *arc*. When two points on the circumference are joined by a straight line, the line is a *chord*, and the part of the circle bounded by a chord, on either side of it, is a *segment*. The area enclosed by an arc and two radii is called a *sector*. Concentric circles have the same centre. The area bounded by two concentric circles is called an *annulus* (see figure 1.9). Figure 1.10 shows some angles and their definitions.

A line (that is, a straight line or a curve) touching another line is a *tangent* — see figure 1.11. A tangent to a curve is a line that passes through two points on the curve infinitely close together.

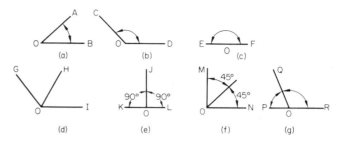

*Figure 1.10   Angle definitions; (a) acute angle, AÔB less than 90°; (b) obtuse angle, CÔD greater than 90° but less than 180°; (c) straight angle, EÔF equals 180°; (d) adjacent angles, GÔH and HÔI lie on either side of common arm; (e) right angles, JÔK and JÔL are equal and adjacent right angles; (f) complementary angles consisting of two angles making 90°; (g) supplementary angles consisting of two (PÔQ and QÔR) angles making 180°. Units of angular measurement are degrees; 60 seconds = 1 minute, 60 minutes = 1 degree, 90 degrees = 1 right angle*

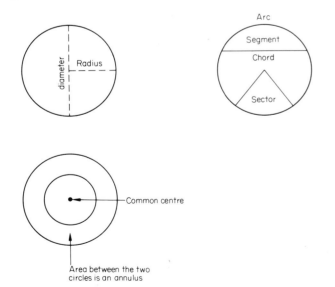

*Figure 1.9   The circle*

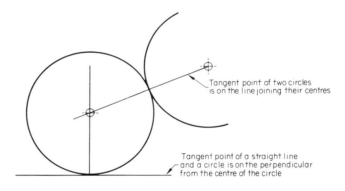

*Figure 1.11   Tangents*

## Area

If any two length measurements are multiplied together the product will be a definition of area and if an area is divided by a dimension of length it will produce the other dimension of length.

## Volume

If three length dimensions are multiplied together the product will represent a volume. Similarly if an area is multiplied by a dimension of length it will produce a volume. Conversely, if a volume is divided by a measure of length it will give an area, and if a volume is divided by an area a dimension of length will result.

## Line Language

The workshop drawing, although employing figures and words to clarify certain points, is basically a language of lines. The thickness, discontinuity or continuity of the lines are used to give the observer a three-dimensional concept of the object, as well as other information. In figure 1.12 the meaning of the lines is illustrated, and figure 1.13 shows some of the symbols used. In addition to using such signs, any other information can be added, either close to a particular section or in the form of a 'note' at the bottom of the drawing.

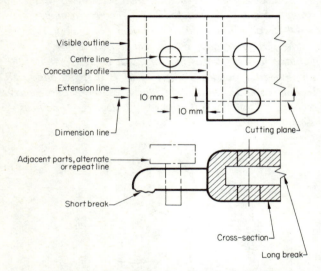

*Figure 1.12    Workshop drawing-techniques*

A grid or working framework of pencil lines is used to start the layout for a drawing, the stages of which are shown in figure 1.14. The isometric view is an optional feature and not normally included in workshop drawings. The section method (see circle A in figure 1.13) is often more explicit and convenient for illustrating details.

Shading or the effective use of lines and colour intensity is often used to suggest roundness. Spheres, radii, tubes, dishes, solid round sections, etc., are often emphasised by shading.

Figure 1.15 shows, in pictorial form, the melting of plates with an oxy-acetylene flame. Precisely what information is presented? What we shall see will depend partly on our prior knowledge. Looking at the welding tip in figure 1.15a we can see that *it is suggested* that there is a source of light on the left. We may infer this because the left side of the tip is not shaded. The lines on either side of the flame inner-cones, in parts (a), (b), (c) and (d), suggest *velocity*. The light shading of the molten areas suggests a *difference* from the surrounding

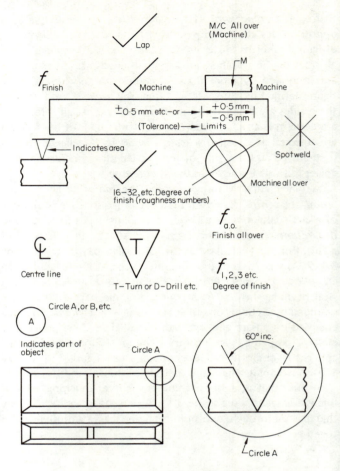

*Figure 1.13    Drawing symbols*

material. Because of our prior knowledge we assume (correctly) that the difference indicated by the shading is thermal.

The sphere of molten metal in figure 1.15a and the hemispheres in figures 1.15 b, c, and d are the visible effects produced on the metal by the flame. The adjacent hemispheres in (b) are for a time separated by the action of the flame and the presence of the vertical discontinuity or joint interface; there is also a condition of surface tension existing on each hemisphere. Eventually, as their diameters increase, their surfaces will touch and fuse together as in figure 1.15c. The condition in figure 1.15d is not quite the same because the weldpool is restricted in size by the boundaries of the plates. The pool will therefore tend to become established faster. So we can now see that in the understanding of a drawing much has to be deduced or imagined, using knowledge previously gained from other sources.

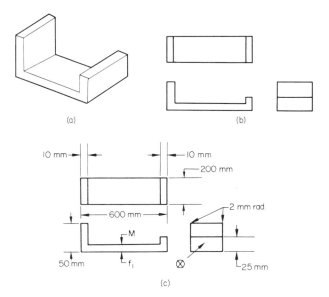

(a)

(b)

10 mm — | — 10 mm
— 200 mm
600 mm
2 mm rad.
M
50 mm
f₁
25 mm

(c)

Figure 1.14   Stages in producing a working drawing; (a) isometric view, (b) locate principal features with light lines, (c) ink in bold outline, add dimensions, machining instructions, etc.

Often a simple isometric drawing such as that shown in figure 1.16 is sufficient to convey enough information. In this case it is only required to show the relationship of the workpiece to a copper block. It is supposed (or left to the observer to realise) that the craftsman, who will eventually weld the joint, knows that there are other important features not illustrated — it would be necessary to clamp or jig the assembly, for instance. The questions arise: What electrodes should be used? Is a welding procedure required? What is the parent material? How thick is the parent material? Should metal-arc or some other process be used? The fact that such information is not given presupposes that the observer has it. Sometimes this is not the case — a good axiom to include on a drawing would be: '*If in doubt, ask*'!

For some purposes a straightforward line-diagram is sufficient. In figure 1.17 there would be no point in making a complicated and expensive drawing when all that it is intended to convey is the type of materials used and the functional aspect of the object to someone only having a remote interest in the subject.

Again, in figure 1.18 the method of producing B from A is all that is implied, the metal that the object is made of, or the method of melting it, is not indicated. The idea of *relationship* is the motive for figure 1.18, the purpose being to show how

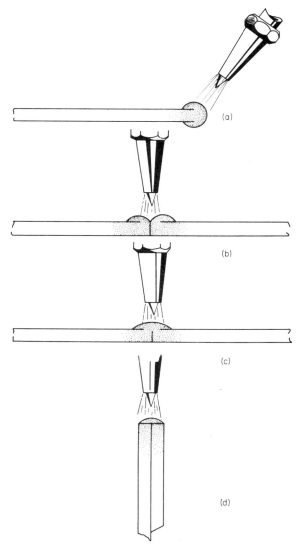

(a)

(b)

(c)

(d)

Figure 1.15   Melting and establishment of weldpool with oxy-acetylene flame

the torch is held, and the position of the fingers and thumb. Shading is again used to suggest roundness and the general dark outline of the hand and torch make the focal points of the drawing.

**Plane Section**

If an orthographic projection is used, hidden features may be complex and confusing, in which case as many sectional views as necessary should be drawn.

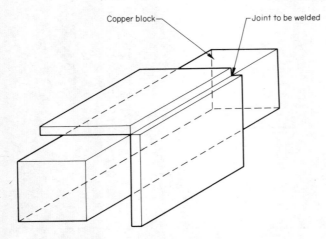

*Figure 1.16    Use of backing bar*

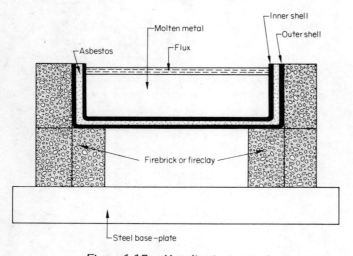

*Figure 1.17    Hot-dip tinning-tank*

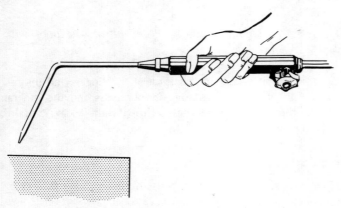

*Figure 1.18    Oxy-acetylene torch*

## Half-section

Usually components that are symmetrical about a centre line (for example, a pipe and flange) are illustrated with this method. Hidden contours are only included if they are necessary for greater clarity or dimensioning.

## Part Section

It may not be necessary to show the full cross-section of a component if some local detail only requires illustration. A bar containing a tapered blind hole at one end, for example, could be shown in this manner.

## Revolved Section

Views of the object are shown by turning it through 90° so that a cross-sectional view may not be required.

## Offset Section

Because it is not always possible to show the desired features of a component by using a single cutting-plane, an offset section may be required. Here the line of cut is arranged to bisect the required member or members of the component.

In addition to providing workpiece details, a drawing often shows machining and tooling information and if, in so doing, more accuracy and detail are conveyed this is commendable. Conversely, too much irrelevant information tends to confuse — *keep it clear and simple*.

# 2. FLUX—METAL-ARC WELDING

Metal-arc welding is a process that employs electricity to generate heat by means of an *arc* between the workpiece and the welding electrode.

## ENERGY SUPPLY

There are three ways of providing the electrical energy: the rotary generator which provides direct current, the rectifier which also supplies direct current and the transformer which has an alternating-current output.

With a d.c. supply, polarity can be changed, that is, the workpiece or electrode can be made positive or negative, as required. About two-thirds of the thermal energy of the arc is generated at the positive pole. Metals requiring a high heat-input, that is, metals having high thermal-conductivity, like copper and its alloys, can best be welded by reversed polarity thus making the work positive. This applies mainly to T.I.G. (tungsten—inert-gas) or 'argon-arc' welding, since metal-arc welding is seldom used today for the welding of these metals. Stainless steels (alloys of iron with mainly chromium and nickel) can be metal-arc welded in the heavier gauges. Electrode polarity is preferably positive because these alloys are poor thermal-conductors. Alternating current can also be used for this purpose.

In a welding circuit the three essential connections are

1. the welding lead or cable connecting the holder to the power source
2. the return cable that provides the means for the current to flow back to the power source
3. the earth connection – this must be of low impedence to prevent any excessive rise of potential in the workpiece or surrounding metal.

The welding return-cable must also be of sufficient section to avoid excessive heating by resistance. This is especially important where more than one welder is working on an assembly using a multi-operator welding-machine.

Some of the hazards arising from unsatisfactory earthing of welding machines are : fire risk through ignition of inflammable materials, short-circuiting and burning out of portable tools, electric-shock risk to operators and damage to apparatus by resistance heating. High resistance in the welding return-cable (caused, for example, by a loose connection) is not only potentially dangerous because it may ignite nearby material, but also causes interruption of the welding current-supply. This directly affects work quality, that is, deposit shape, penetration, slag control, arc striking, etc.

### The Electrode Holder

This is basically a particularly robust pair of spring-grip pliers with provision for the connection of the welding cable and constructed usually of a copper alloy. All parts of the holder, except the inner faces that grip the electrode, should be insulated. It should be remembered that the holder is 'live' when the machine is switched on. Do not leave the holder on the bench, in contact with combustible material or near portable electric apparatus. Short-circuiting of the welding current in such circumstances is dangerous. All connecting leads must be of the recommemded type and of adequate conductive capacity — odd pieces of metal should never be used to make up a welding circuit.

## SAFETY

Besides generating heat the electric arc also emits radiation in the visible and invisible spectrum. Infrared and ultraviolet radiation are harmful to the eyes and unprotected areas of the body. The condition known as 'arc eye' or 'the flash' is due to exposure of the eyes to an electric arc. Students handling arc-welding apparatus should take special care to prevent this happening. Striking the arc for the first few times will, for example, be difficult and the natural tendency is to move the welding mask away from your face to see what is wrong. Often the electrode has 'frozen' to the workpiece, and because the electrode starts to get hot, by resistance, the student may panic and try to wrench it free. Doing so without the protection of the mask, exposes your face at close quarters to the arc. Arc eye although painful is rarely permanently damaging but if exposure is frequent, it may cause prolonged conjunctivitis. Cold-water packs and rest are necessary for the treatment of arc eye and medical attention must be sought. The welding mask should be provided with the correct shade of filter glass, which is protected by plain glass on both sides.

### Ventilation

The fumes generated by the metal-arc welding of iron and steel are not toxic, but protective coatings (for example, galvanised, painted, nickel plated, etc.) may produce a toxicity risk. Welding should therefore be conducted in a well-ventilated environment, otherwise local fume-extraction should be provided. Always keep the welding booth clear of obstacles and potential hazards such as cylinders of compressed gas, empty or full drums, inflammable materials, scrap metal, water and items of clothing, etc.

### Clothing

Special protective clothing should be worn for welding, as recommended in BS 1542:1960 Equipment for eye, face and neck protection against radiation arising during welding and similar operations. White coats may promote arc eye because they reflect radiation; black, navy blue, brown and dark colours absorb it.

Loose clothing (shirt cuffs, long cotton overall coats, etc.) is dangerous — use only the appropriate clothing. Clear glass goggles with shatter-proof glass or plastic lenses should be used when chipping slag from welds. Burns to the eyes either by hot slag or radiation are very painful and you should therefore always wear goggles for chipping — learn the easy way!

### ELECTRODES

### Electrode Coatings

The fusion joining of metals using electrodes coated with flux is known as metal-arc welding. It can be used manually as shown in figure 2.1 — note the use of magnets holding the workpiece — or automatically as shown in figure 2.2; the flexible coiled electrode used here consists of several wires twisted to form a helix, the interstices of which are filled with flux, this is known as the fusearc process.

An essential feature of good general-purpose electrodes for metal-arc welding is their capability of operation in any position — flat, vertical, overhead, inclined, horizontal—vertical, etc. Metal is transferred across the arc in modes ranging from spray to globular, depending on current and electrode type. For example, rutile- or titanium-oxide-coated electrodes at normal currents exhibit a globular metal transfer, with droplet diameters from about 1.9 to 2.3 mm. When an arc has been struck, the electrode tip begins to heat up rapidly and the surrounding flux melts and flows towards the arc root. Simultaneously, flux decomposition proceeds and thereby evolves carbon monoxide and other gases such as water vapour, nitrogen, etc., in small amounts, these products being drawn into, and around, the arc. The metal globules streaming from the electrode are coated with flux which becomes the 'slag' overlay, covering the weld metal.

Electrodes are classified under BS 1719:Welding electrodes: Part 1:1969 Classification and coding and BS 639:1972 Covered electrodes for the manual metal-arc welding of mild steel and medium-tensile steel, and range in diameter from 1.6 to 10 mm. The old method of manufacturing electrodes, by repeatedly dipping the core wires into liquid flux and allowing them to dry, has been superseded by the coating-extrusion process. Here the wire is fed continuously through a die into which flux is forced at high pressure.

Fluxes consist of binding agents like gums, clay, sodium silicate and ionising materials such as titania (titanium oxide), potassium silicate, aluminium silicate and sodium salts. A reducing atmosphere — one containing hydrogen, for example — is provided by using wood flour. This is the cellulosic type of electrode that is very popular for general-purpose production since it gives a forceful arc and a light slag coating. Calcium fluoride and alkaline metal carbonates are added to produce basic-slag-type electrodes (low-hydrogen). Finely divided iron is used in some electrode coatings to give a higher than normal metal deposition and recovery rate. In some cases the low-hydrogen characteristic is combined with the iron-powder one in an attempt to obtain the best features of both additives.

Asbestos and mineral silicates and oxides of manganese and iron are used to give properties of light, inflated or dense slag. These characteristics have an important bearing on positional welding, thus when welding vertically the slag must freeze quickly and also provide a substantial lip to the weld crater to support the molten pool. Deoxidizing compounds used in electrode coatings, consist of ferromanganese and silicon in the form of ferrosilicon or silica. By adjusting the various flux components, a great variety of arc and deposition characteristics can be provided.

*Figure 2.1    Magnetic clamping (courtesy of James Neill and Co. Ltd)*

**Functions of Electrode Coatings**

Bare wire electrodes do not produce satisfactory welds because the liquid deposit is attacked by atmospheric gases such as oxygen, hydrogen (from water vapour) and nitrogen. To counteract these effects, electrodes are coated with flux, which decomposes in the arc to generate a protective gas, usually carbon dioxide. Additionally, the flux provides ionising constituents which improve electrical conduction through the arc. Slag-forming materials are also incorporated, which have a scavenging or refining action on the molten metal. Another function of flux is to control bead (deposit) shape. Alloying metals such as manganese, chromium, nickel, molybdenum and vanadium are often included in fluxes for the welding of special steels. Fluxes are often hygroscopic, that is, they absorb water vapour. This promotes chemical changes in the flux besides providing an unwanted source of hydrogen and thus impairing weld quality. Electrodes should therefore always be kept dry; if allowed to become damp they can be dried in an electrode oven provided with a temperature control

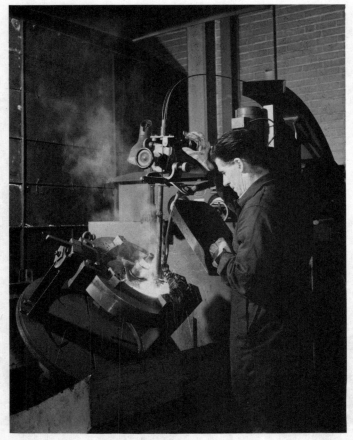

*Figure 2.2    Fusearc process — brake-shoe welding (courtesy of B.O.C. Ltd)*

(thermostat). Among the other reasons for coating electrodes are: introduction of alloy materials, influencing metal transfer across the arc, increasing deposition rates and determining slag properties.

## Electrode Types

In most situations, a short arc-length is employed with an electrode since a long arc increases arc voltage and thermal-energy loss by radiation. Metal transfer becomes increasingly erratic and the arc root (point of impingement on the workpiece) tends to wander. Electrode types and characteristics cover a wide range and are often designed for use under special conditions or with particular alloys. For example, in the 'stovepipe' technique for pipeline welding, an electrode producing voluminous slag, taking a comparatively long time

to solidify, would be unsuitable. This is because this welding application uses the vertical down technique calling for light quick-freezing slag.

Alloy metals must be welded with electrodes containing the same constituent metals; stainless steel, for example, is welded with electrodes containing compensating amounts of chromium, nickel, etc., in the core wire or flux.

High-tensile steels containing such elements as carbon and manganese are often welded with electrodes known as 'low-hydrogen' types. These are coated with material containing no chemically combined hydrogen but up to 18 per cent calcium flouride with calcium carbonate. Hydrogen is particularly undesirable in this case because it can produce microcracking with subsequent component failure.

Electrodes, generally, have been developed from two main types, mineral and organic. Mineral-coated electrodes can be divided into the basic-slag and acid types. Low-hydrogen electrodes belong to the basic-slag category. High-sulphur and free-machining steels can also be welded with these electrodes. Electrode coatings designated low-hydrogen are classified under class 6 — electrodes having a high proportion of calcium carbonate, mineral silicates and calcium flouride.

Among other types of electrodes for the welding of steel are the cellulose type — class 1 — containing about 30 per cent titania (titanium oxide) with up to 15 per cent cellulose and minor additions of ferromanganese and silica, both of which are deoxidisers.

Class 2 electrodes are of the 50 per cent titania type. This provides a quiet stable arc-condition and a viscous slag.

Class 3 electrodes, in addition to titania, contain calcium fluoride which further increases slag fluidity.

Class 4 electrodes are not so common in use; they produce voluminous slag and a concave weld-deposit; they are limited in use to flat or down-hand welding.

Class 5 electrodes are those with coatings having a high iron-oxide content with silicates. These electrodes are not often used because the weld metal is oxidised and therefore not compatible with most quality standards.

Another special electrode for use with steel is the iron-powder type, containing from 10 to 50 per cent by weight of finely divided iron in the coating. This increases the deposition of weld metal per electrode as well as the speed of welding, because of the enhanced conductivity.

There are many electrodes for special materials and purposes, some of these are for depositing hard alloys, welding cast iron, stainless steels, etc., high-tensile steels, some copper alloys (now mostly T.I.G. welded) and, decreasingly, heavy aluminium sections (also now mainly T.I.G. welded).

### Electrode Storage

Since many electrode coatings are hygroscopic — that is, they readily absorb water or water vapour — they should always be preserved in dry conditions. This is especially important for low-hydrogen types because their main function is to provide a weld free from hydrogen inclusions. Water picked up by an electrode decomposes in the arc and the hydrogen evolved dissolves in the molten metal.

### POWER SOURCES

There are three types of power source for arc welding: the transformer, the motor generator and the a.c./d.c. transformer—rectifier (see figure 2.3). The transformer is most

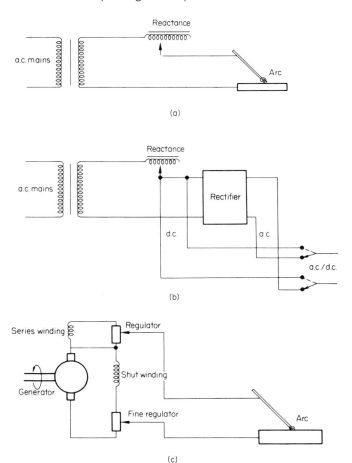

*Figure 2.3   Power sources; (a) a.c. transformer (single phase), (b) a.c./d.c. power source (single phase), (c) d.c. motor generator*

commonly used for general work because of its low capital and operating costs.

With an a.c. supply any of the three power-sources may be used but the motor generator is most often used for site work because as an alternative to being driven by an a.c. motor, a diesel or petrol engine may be used.

When using a transformer, the output is regulated by a reactor or choke which alters the value of the secondary-circuit current. The rectifier unit consists of a transformer whose output is connected to a metal rectifier to provide a d.c. output. This power source is also used for $CO_2$ (carbon dioxide) welding.

If a single-phase input only is used the output current is not smooth d.c.; to obtain this it is necessary to use a three-phase supply. Whereas the open-circuit voltage required with a welding transformer is from 80 to 100 V, because of the a.c. characteristic which causes periods of arc extinction, 40 to 60 V is sufficient with d.c. equipment. For metal-arc welding, but not for $CO_2$ welding, d.c. power sources must have a drooping characteristic, which means that large changes in arc voltage only cause small changes in current — see curve A in figure 2.4. This is because in manual arc welding, arc-length variation is inevitable, although in certain instances such as overhead or vertical welding, it is an advantage to be able to change current by altering the arc length, producing a curve as in curve B.

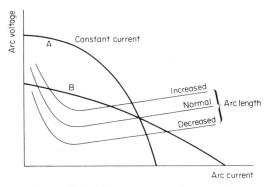

*Figure 2.4    Power-source characteristics*

Although the open-circuit voltage must be high enough to provide easy arc striking, there are exceptions to the 100 V upper limit. In this case special precautions must be taken to protect the operator from accidental contact.

### Electrical Conductors

An electric current flowing through a conductor creates a magnetic field around it and may also generate heat owing to the conductor's resistance to the passage of current. It is not essential to use metal for electrical conduction — certain liquids, ionised gases and non-conductors (when strongly heated) may be used. For example, the minerals silica and cryolite (an aluminium ore) become conducting when molten. Certain elements like carbon and selenium are only conductors in specific states. Carbon in the form of graphite, for instance, is normally a good conductor (hence its use as brushes in electric motors, etc.) but diamond, that is, crystalline carbon, is an insulator.

Gases or mixtures of gases, like air, which when dry are good insulators, may be made to conduct an electric current by ionisation. The best example of this is a thunderstorm, where leader strokes of high voltage precede the passage of lightning. In the case of the welding arc, ionisation of the air is achieved by touching, or short-circuiting, the electrode to the workpiece to which is connected the other terminal of the power source. When the electrode is withdrawn an ionised gap or arc is established — see figure 2.5.

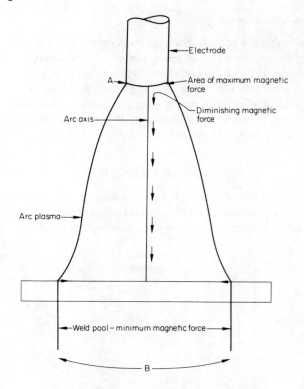

*Figure 2.6    Physical structure of the arc*

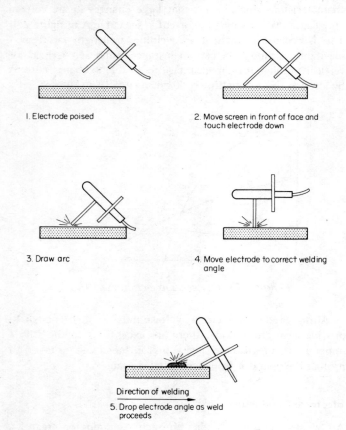

*Figure 2.5    Striking an arc*

The magnetic field generated by the flow of current through the electrode also encompasses the arc, subjecting it to a constricting or squeezing force. The current flowing across the ionised arc-gap is therefore also flowing through its own magnetic field, so that the electrons streaming from the electrode are subjected to a force directed towards the longitudinal axis of the arc — see figure 2.6. The small radius, A, represents the electrode tip, where the constricting force is greatest; the force diminishes towards the weldpool, B. The arc can be constricted by mechanical means — this principle is used in the generation of the plasma arc. Figure 2.7a shows a normal transferred arc between electrode and work. Figure 2.7b shows a constricted transferred arc. Figure 2.7c shows a non-transferred arc between an internal and external electrode and figure 2.7d shows a non-transferred arc with both electrodes inside the torch. The length of the non-transferred arcs in figures 2.7c and d can be increased by increasing plasma-gas and shielding-gas pressure. In all three cases, b, c and d, one electrode is tungsten, protected by a gas shield of argon, helium or a mixture of these or other gases.

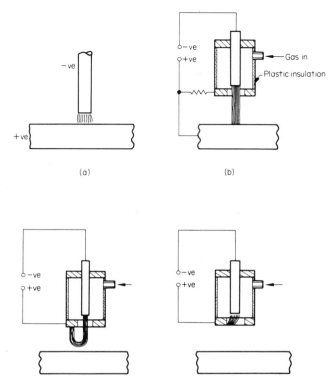

*Figure 2.7 Arc types; (a) open arc, (b) transferred-arc torch, (c) non-transferred arc, exterior arc-contact, (d) non-transferred arc, interior arc-contact*

## CARBON-ARC WELDING

Carbon-arc welding is little used today but is still a good introductory technique for the student. A direct-current power-supply is required, the carbon electrode being connected to the negative output-terminal. About two-thirds of the available heat energy of an arc is generated at the positive pole and this is used to advantage not only in carbon-arc welding, but in metal-arc, tungsten—inert-gas (T.I.G.) metal—inert-gas (M.I.G. or $CO_2$), plasma-arc and submerged-arc welding.

Figure 2.8 illustrates the principle of carbon-arc welding, the filler wire being kept (as in manual T.I.G. welding) near, or within, the perimeter of the arc. Carbon monoxide and carbon dioxide are generated by the carbon arc and these help to protect the weldpool from atmospheric attack. For welding light-gauge material the arc should be kept short. Carbon (or

*Table 2.1 Typical conditions for carbon-arc welding*

| Current (A) | Voltage (V) | Electrode diameter (mm) | Electrode diameter (in.) | Arc length (mm) | Arc length (in.) |
|---|---|---|---|---|---|
| 75 | 60—90 | 5.0 | 3/16 | 19.0 | 3/4 |
| 100 | 60—90 | 6.5 | 1/4 | 19.0 | 3/4 |
| 200 | 60—90 | 9.5 | 3/8 | 19.0 | 3/4 |
| 300 | 60—90 | 12.5 | 1/2 | 16.0 | 1/2 |
| 400 | 60—90 | 19.0 | 3/4 | 16.0 | 5/8 |
| 600 | 60—90 | 25.0 | 1 | 25.0 | 1 |

graphite) electrodes are fragile and cannot be used with the standard type of metal-arc-welding holder — a special holder is required which uses a very light spring. Table 2.1 shows typical conditions for carbon-arc welding.

An alternative method of carbon-arc welding is shown in figure 2.9 where two electrodes are held next to each other by means of an adjustable insulated yoke. In this system the workpiece need not be connected to the power source but it is still good practice to do so, to avoid possible damage to other equipment by short-circuiting. A carbon electrode protected by an argon shield can be used for depositing aluminium bronze, especially on galvanised steel. The carbon arc in argon is quiet and has the advantage over conventional T.I.G. that metal cannot stick to the electrode, whereas a tungsten electrode used for arc brazing can become contaminated by globules of metal sticking to it and causing arc instability.

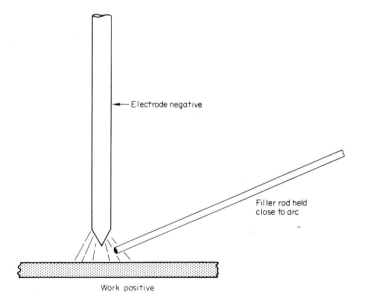

*Figure 2.8 Carbon-arc welding*

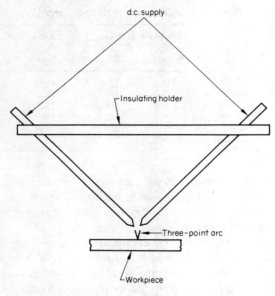

Figure 2.9    The electric 'blowpipe'

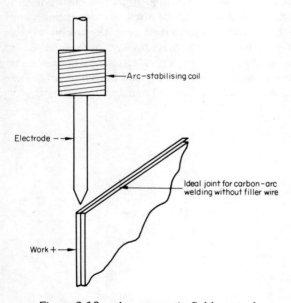

Figure 2.10    Arc magnetic-field control

For automatic welding the electrode holder should be fitted with an arc-stabilising coil — see figure 2.10. This may have a separate electrical supply, but for manual use the power-source output can be tapped. The edge joint is ideal for carbon-arc welding, for which a separate flux, or flux-coated filler-rods, may be used. Tacking edge-to-edge joints in light-gauge steel

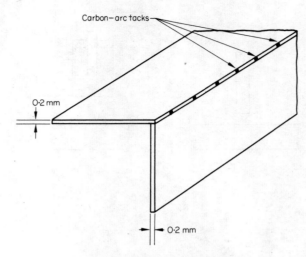

Figure 2.11    Use of carbon arc

with the carbon arc (see figure 2.11) largely eliminates the distortion that would occur if an oxy-acetylene torch were used. Another advantage is that because the tacks are very small, especially if made without using filler material, the subsequent weld has a smooth even contour.

The carbon arc can also be used for welding cast steel, cast iron, stainless steels and brazing. Butt edge-joints can be made in material from about 0.8 to 3.2 mm thick, with or without the addition of a filler wire. The maximum welding-speeds possible with carbon arc are

Manual — 37 m/h
Machine welding with an automatically fed electrode — 120 m/h

## SOLUBILITY OF GASES IN METALS

Since molten iron has a greater capacity for hydrogen adsorbtion than the solid metal, the excess hydrogen is partly given off during weld cooling. Not all of it may escape, the residue being present in the form of *voids* which are really trapped bubbles. Oxygen is also soluble in iron, reacting with carbon to form carbon monoxide, which also comes out of solution as the metal cools, resulting in porosity. Molten iron is also a solvent for nitrogen and, as a result, hardening may occur due to iron-nitride formation. Table 2.2 compares gas solubility in iron.

Hydrogen is soluble in liquid aluminium and copper and if the latter contains any copper oxide when heated in contact with a hydrogen source, water vapour may be produced and

Table 2.2    Comparative solubility of
nitrogen, hydrogen and oxygen in iron

|  | Solubility (% by weight) | |
|---|---|---|
|  | at 1550°C | at ambient temp. |
| Oxygen | 0.16 | 0.0 |
| Nitrogen | 0.040 | 0.0001 |
| Hydrogen | 0.0025 | 0.0005 |

gross weld-porosity will result. The inert gases argon and helium, used in T.I.G. and the plasma-arc processes, are insoluble in most metals. Nitrogen is not soluble in copper. Oxygen and nitrogen are not soluble in aluminium, but special care must be taken when welding titanium and its alloys, because hydrogen, oxygen and nitrogen are all soluble in them, and will cause severe embrittlement.

## PREPARATION OF WORKPIECE

It is most important for the execution of high-quality welded work that joint preparation and type are correct. Time spent in work preparation is not wasted but provides a quality bonus.

Joint design is dictated by

1. thickness of material
2. type of material
3. position in which the work is to be done (for example, flat, horizontal, inclined, overhead, etc.)
4. type of welding process available
5. degree and conditions of presentation or access to work by the welder (*this point, unfortunately, is often overlooked*)
6. service conditions of the welded structure.

The normal welding-positions are designated as flat (f) or 'downhand', inclined (i), horizontal (h), vertical (v), overhead (o) and horizontal—vertical (h.v.). Figure 2.12 shows some commonly used joints for metal-arc welding and in figures 2.13 and 2.14 examples are given of joint measurements and coding.

The butt joint has many variations, from the closed butt-joint — consisting of the edges of two plates in contact — to the single or double vee, 'J' and 'U' preparations that are used for heavy-gauge material. The closed butt and open-square butt (provided with a gap between the joint faces) are used a great deal in light-gauge material fabrication and are suitable for metal-arc, oxy-acetylene, manual or automatic T.I.G., M.I.G. or plasma-arc welding.

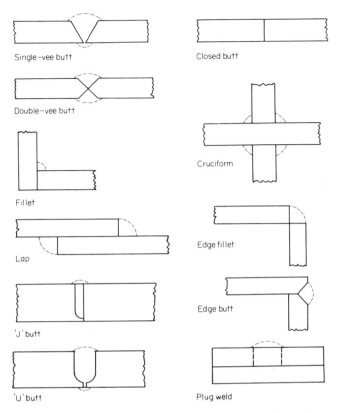

Figure 2.12    Joint preparation; for full details refer to BS 499: Part 2: 1965 Symbols for welding

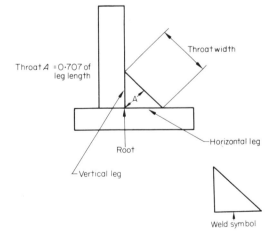

Figure 2.13    Fillet dimensions; for full details of joint design and symbols see BS 499: Part 2: 1965 Symbols for welding, and for further reference see American Welding Society Standard A20—58

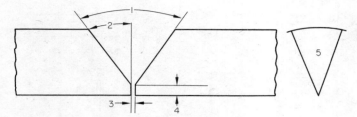

*Figure 2.14    Single-vee joint with root gap and root face;*
*1 — included angle, 2 — angle of bevel, 3 — root, gap, 4 — root*
*face, 5 — weld joint symbol*

For metal thicknesses above 8 mm the open butt can be welded using a backing bar, which may be of copper and merely retain the molten metal, or it may be of the same material as the weldment and subsequently form part of it.

Figure 2.15 shows some typical edge preparations for a variety of metal thicknesses: (a) is suitable for metal-arc, T.I.G. or M.I.G. welding; (b) is used for thick-to-thin joints, both (a) and (b) being ideally welded in the flat or downhand position; (c) is mostly employed in sheet-metal fabrication using metal-arc, oxy-acetylene, T.I.G. or M.I.G. in all posit-

ions; (d), (e) and (f) in heavy-gauge material are suitable for welding by metal-arc and submerged-arc; (g) is the most economical edge joint and is used in light-gauge-material fabrication with metal-arc, oxy-acetylene, carbon-arc, or T.I.G., but for this joint (in metal gauge combinations of less than 2 x 2 mm thickness) M.I.G. welding is not satisfactory especially if a neat finished appearance is required — in this respect oxy-acetylene or T.I.G., without added filler metal, produce the best results. The joint shown at (h) is to be preferred to (i) which offers poor conditions of access, with the resulting risk of slag inclusions and lack of root penetration at A—A; (j) shows the way in which the situation may be improved.

The mismatch butt-joint in figure 2.16a is unsatisfactory, and if one or both horizontal surfaces are subsequently machined flat the joint strength will be considerably reduced. The mismatch in figure 2.16b can be produced by shearing the edges on a guillotine with damaged cutting-edges and, if automatic welding is employed, this will cause penetration inconsistencies. The over-reinforcement in (c) means extra finishing-costs, waste of material, electricity and labour, and should be avoided. Figure 2.16d shows undercutting, which in

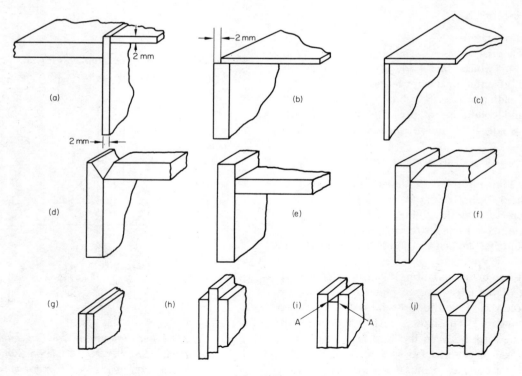

*Figure 2.15    Typical edge preparations*

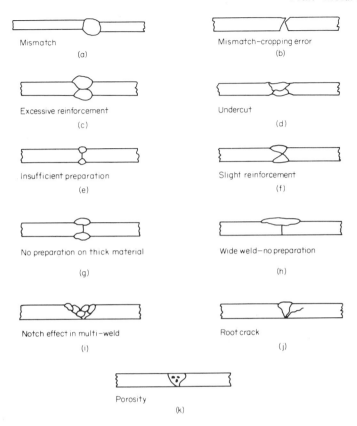

Mismatch
(a)

Mismatch—cropping error
(b)

Excessive reinforcement
(c)

Undercut
(d)

Insufficient preparation
(e)

Slight reinforcement
(f)

No preparation on thick material
(g)

Wide weld—no preparation
(h)

Notch effect in multi–weld
(i)

Root crack
(j)

Porosity
(k)

*Figure 2.16    Butt joints*

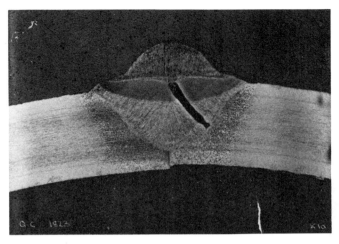

*Figure 2.17    Piping porosity*

*Figure 2.18    Off-line weld*

most cases may be rectified by rewelding. Figure 2.16e shows poor preparation, but (f) is satisfactory because it has minimum reinforcement without undercutting, cracking or porosity and shows complete penetration. Figures 2.16g and h show faults that may arise by attempting to make butt joints in thick material without preparation, while (i) illustrates an undesirable notch caused by insufficient filling of the joint. Joints, as shown in (j) and (k), contain internal defects that may sometimes also appear at the weld surface. Some degree of porosity can be tolerated, depending on required standards, but cracked welds should *never* be tolerated. In some instances, the whole of the weld deposit may have to be removed by chiselling, oxy-acetylene or carbon-arc—air gouging or machining; but minor cracks (for example, surface cracks of limited length and shallow depth) can often be dealt with satisfactorily by local removal and rewelding.

Good joint design and meticulous preparation go together with correct welding-procedure in producing high-quality work — *there is no short cut*

Piping porosity, so-called because of its resemblance to a tube, can be seen in figure 2.17. Because the fault had not been removed, the ineffectiveness of the subsequent weld-deposit can clearly be seen; the location and extent of the piping strikingly illustrates the seriousness of this type of defect which, in this case, was due to hydrogen pick-up (the slight mismatch should also be noted).

Figure 2.18 shows a weld having incomplete penetration at the root, slight undercutting of the top surface and mismatch of the parts. Failure in the hardened zone of the weld is seen in figure 2.19 which also shows a lack of penetration.

Certain joint configurations — see figure 2.20 — should be avoided if possible, although there are times when exceptions

*Figure 2.19    Lack of root fusion*

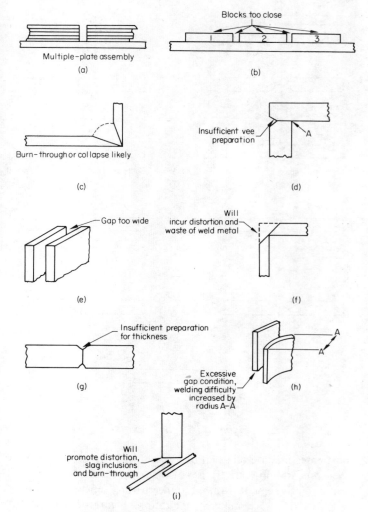

*Figure 2.20    Unsatisfactory assembly*

have to be made. The multiple-plate assembly (figure 2.20a) is not recommended because of the limited access provided and the inherent risk of poor root and side-wall fusion and slag trapping. In figure 2.20b the members 1, 2 and 3 are situated too closely together, a similar condition to (a). To fill the joint in figure 2.20c an excessive deposit of weld metal would be required and there is the great probability of melting away the sections with the consequent loss of the weldpool. Figure 2.20d shows insufficient preparation, and on thick material a fillet weld at A would not be likely to compensate by adequate joint penetration. Figure 2.20e illustrates a common fault, that of excessive gap between the plates — an unsatisfactory condition in any gauge of material. In figure 2.20f the large weld-deposit necessary would promote severe distortion of the workpiece. Insufficient preparation in figure 2.20g would mean unsatisfactory weld penetration, in the case of heavy material; the gap condition in figure 2.20h is even worse than that shown in (e) because of the radius A—A that provides an increasing gap condition.

The joint in figure 2.21a is unsatisfactory for metal-arc welding because of the danger of incomplete penetration and root fusion; in addition, slag could easily be trapped at the root. In figure 2.21b the side members $B_1$ and $B_2$ would collapse if the joint were filled in a continuous operation. In figure 2.21c a condition similar to (a) is seen but with the additional probability of overheating and collapse.

Welding round-section material to flat plates never gives a good welding-condition, especially in larger diameters, because of poor root-access and the risk of undercutting on the round

component — see figure 2.21d. The gap condition in figure 2.21e together with the large sectional difference of the parts, makes this unsuitable for welding, but, as in figure 2.21f, a backing bar could be provided. Access to the joint faces in figure 2.21g would be so difficult that efficient welding would be impossible, besides which the electrode would almost completely obscure the welder's view of the work area. The narrow angle provided by the joint in figure 2.21h together with the work position would provide conditions for poor root-fusion, slag trapping and undercutting. The joint in figure 2.21i would be difficult to weld without backing, while that in figure 2.21k would require an excessive amount of weld metal and produce severe distortion of the workpiece. Single-vee

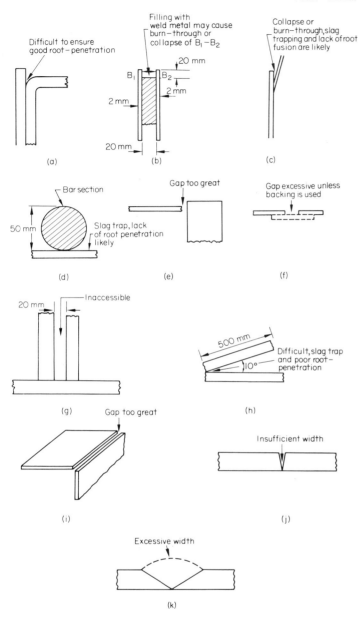

*Figure 2.21    Unsatisfactory joints*

## PENETRATION

Complete joint-penetration can be achieved with no root opening or preparation (the square closed butt) on material up to about 5 mm thickness by metal-arc welding. On sections up to 10 mm thick a single- or double-vee preparation with or without root face and root gap is required. Thicker material can be prepared with a double vee, 'J', 'U', or a combination of these, but in this range, except for site constructional work, it is not usual to employ the metal-arc process. Automatic submerged-arc, electro-slag or continuous flux-filled tubular-electrode welding may be used. This latter process employs an electrode made from thin steel strip and formed into a tube filled with flux. Very high deposition-rates are possible in the flat position and irregularities of joint fit-up are more easily accommodated than with M.I.G. or submerged arc. The process is called flux-cored-electrode welding or 'innershield' welding and in some applications may include the provision of a carbon-dioxide gas-shield.

The metal-arc process is less efficient than T.I.G. or submerged-arc welding as regards the conversion of electrical energy into thermal energy because of the need to use part of the available power to melt the flux, and also because of the inadequate resistance-heating of the electrode; for example, the electrode diameters used in metal-arc welding are usually so large that cathodic or anodic heating becomes the primary source of thermal-energy emission. Since the current that can be used with a covered electrode is related to its diameter, increase in current above a specific electrode capacity means an increase in diameter, whereas with T.I.G. or submerged-arc welding there is much greater latitude for current increase for a given electrode size.

## THE CRATER

The depression or crater is the region in which the welding arc impinges on the workpiece. Considering it another way, it is the opposite pole — anode or cathode, as the case may be — to the electrode.

Figure 2.22 illustrates the comparative positions of the electrode and the arc route taken to finish off the weld. During the execution of a weld by the metal-arc process, the movements necessary are

1. striking and establishing an arc
2. moving the electrode to the correct welding-angle
3. moving the electrode slightly to the left of joint centre line

preparations are usually 60 or 70° inclusive, except for special metals and conditions, for example, cast iron for oxy-acetylene welding in thick section may require an 80° angle vee. The opposite condition is shown in figure 2.21j – this narrow preparation would mean insufficient root-fusion, and slag trapping.

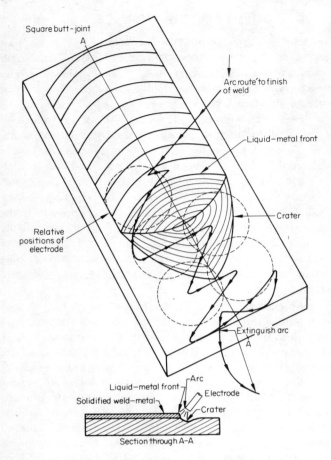

*Figure 2.22     Electrode manipulation*

4. moving the electrode slightly to the right of joint centre line

5. simultaneously with 3 and 4 moving the electrode along the joint and gradually increasing the electrode vertical alignment to the work thus conpensating for the electrode shortening.

The crater that is formed almost immediately the arc is struck remains while metal is being deposited and accumulated behind it. Careful observation of the crater during welding is necessary because its position relative to the joint centre line determines the accuracy of weld deposition. The perimeter of the crater bounded by the molten metal must be carefully observed to impose slag control — a factor largely determined by the way in which the electrode is manipulated. Too much bias, either to left or right of the centre line, may allow slag to overrun the crater causing an inclusion to be made in the

finished weld. An arc that is erratic for any reason, such as poor electrical-return, wrong electrode-angle or manipulation, etc., will also encourage a slag invasion of the crater area. A skilled welder, assuming his welding parameters are correct, may rectify the condition by quickly shortening the arc, thus forcing the slag back over the crater edge. Although the presence of the crater has to be accepted during deposition, it should not be left unfilled on completion of the weld.

Craters occurring on the edges of plates, in corners, or at the end of tack welds, etc., constitute high-risk cracking-areas and for this reason they should always be filled up to the level of the weld deposit. One way in which this can be done is shown in figure 2.22 where it can be seen that the arc route alters obliquely towards the end of the weld. This is done to retain the advance of the molten metal towards the plate edges, the arc finally being withdrawn slightly to one side of centre. At this point the electrode should be withdrawn sharply so as not to promote breakdown and melting of the plate edges.

Another way to finish a weld is to reverse the direction of electrode travel along the joint at a point beginning at about 15 mm from the end. By doing this the electrode is moved backwards over the deposit then immediately withdrawn. Another alternative consists of finishing the weld leaving the crater, then restriking the arc to fill the crater.

## WELDING PROCEDURES

Neatness and consistency in welding are important factors, not necessarily because of aesthetic considerations alone, but uniformity of deposition helps to offset stress concentrations arising from excessive build-up of weld metal, craters unfilled at intersections or at sudden changes of section, etc.

Figure 2.23 illustrates a method of avoiding the ending and starting of welds on corners, and figure 2.24 shows a method of achieving a neat and continuous fillet in a corner joint. The 'hot tack' is made first by using substantially more current with the electrode than that required for actual welding; for example, for a 3.25 mm (10 S.W.G.) electrode normally used at 100 A, the short-time hot-track current would be about 140 A. If the electrode were to be used at this abnormally high current for continuous deposition, it would rapidly overheat and collapse, but because the conditions only apply for a short time — about 2 seconds — no harm is done. When the hot tack has been made good fusion into the joint corner should have been achieved because of the high heat-input. To see what the opposite effect would be, try making the tack with the same-gauge electrode at, say, 70 A. After chipping the slag

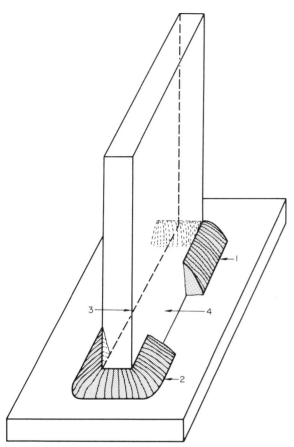

*Figure 2.23    Welding procedure; make welds 1 and 2 around corner before completing with welds 3 and 4*

from the tack, the two main welds, starting from just around the corners as shown, can be made. If a weld at right angles (in the vertical plane) is subsequently required, a neat and slag-free joint can now be made at the intersection of the first two welds without an unsightly build-up of weld metal or slag traps.

A useful procedure for producing neat continuous welds around small components (tapping blocks, or studs) is shown in figure 2.25.

1. A small tack is made.
2. The alignment of the block is now checked because the alternate expansion and contraction occurring when the tack was being put down, amy have pulled it out of position.
3. The corner opposite the tack is now tapped down or, preferably, clamped.

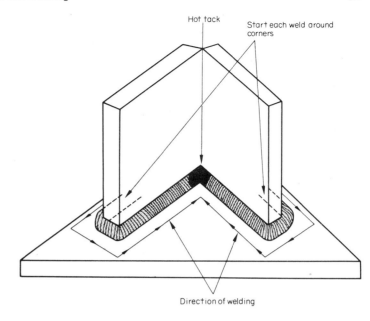

*Figure 2.24    Welding procedure*

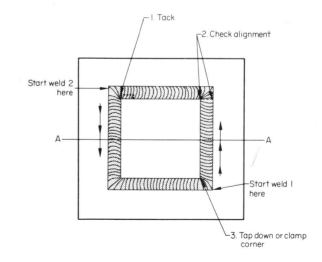

*Figure 2.25    Welding procedure*

Weld 1 is now started at this point towards, and joining up with, the tack. Weld 2 commences at the tack and proceeds towards the start of weld 1.

Specified weld dimensions should be observed as closely as possible. On light-gauge material, say from 0.6 to 2 mm thick, oversize fillet-welds greatly increase the prospect of burn-through or parent-metal collapse. Where the fillet is of the correct size and of equal leg-length, as in figure 2.26a, this danger is remote. Here members A and B receive the same amount of heat, but in figure 2.26b most of the weld metal has been deposited on the surface of A, a condition likely to cause collapse. In addition, the effectiveness of the weld has been greatly reduced because of the inequality of leg lengths.

A weld deposited on the edge of a thin sheet, as in figure 2.26c causes it to become grossly overheated in this area and the weldpool is most likely to be lost by the collapse of the sheet edge. If we now suppose that case (c) becomes A in assembly (b), which has a gap at the point of intersection of A and B, it will be seen why the molten metal may be lost during welding. Weld metal deposited some distance from the sheet boundary is less likely to destroy the edge or be lost — see figure 2.26d.

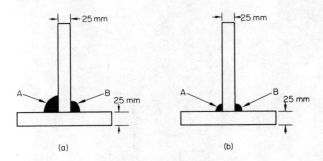

Figure 2.27    Weld size; (a) excessive amount of weld deposit at A causes distortion and may promote cracking in weld B, (b) A and B 4 mm welds — these are too small in relation to plate thickness and are liable to crack

On heavier-gauge material (see figure 2.27a), while burn-through is far less probable, distortion due to incorrect weld-disposition can be severe and often costly to rectify. Welds that are too small in relation to plate thickness, as in figure 2.27b, are liable to failure because they will have inadequate root-penetration and will also be impaired by the rapid-quench effect brought about by the heavier plates.

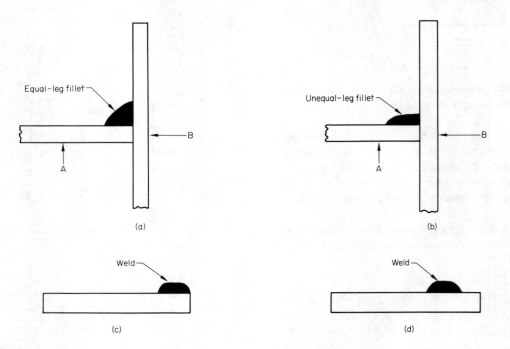

Figure 2.26    Welding technique; (a) heat input to A and B equal, (b) preferential heat input to A promoting burnthrough, (c) weld causing collapse of edge, (d) weld away from edge prevents edge melting

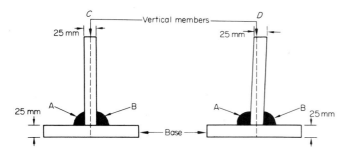

*Figure 2.28   Correct weld dimensions; (a) A and B made simultaneously, keeps vertical members square with base plate, (b) A or B made consecutively causes misalignment of vertical members*

One way in which distortion of the vertical member of an assembly can be greatly modified is by adopting the procedure shown in figure 2.28a. Here both fillet welds are made simultaneously and in the same direction. Some distortion of the horizontal plate will occur, but in most cases presetting by clamping will substantially reduce it. If the welds are made separately and without any restraint of the plates, distortion will occur as in figure 2.28b and will be very difficult, if not impossible, to rectify.

### Tacking

A little thought given to the size, length and location of tack welds substantially reduces the risks of joint misalignment (occurring progressively with the deposition of the weld bead) and root cracking. The latter may happen if the tack weld is unable to withstand the shrinkage stresses set up by an adjacent weld. For example, 5 mm leg-length fillets, 5 mm long and spaced at, say, 150 mm intervals along one side of a 'T' joint, would certainly fail if a continuous 20 mm fillet weld were made on the opposite side.

The properties of the parent material have an important bearing on the way tacks should be disposed, their size and length. Armour plate, for instance, is a high-tensile heat-treatable steel, and distortion set up by welding cannot so easily be rectified as would be the case with mild steel. Armoured fabrications, such as tanks, military equipment, etc., are assembled from plates machined often to close tolerances, one reason being that the increase in hardness adjacent to welds makes further machining very difficult. Welding procedures therefore must be strictly observed including, of course, pre-heating post-heating and tacking. No hard and fast ruling can be laid down for tacking, since many other

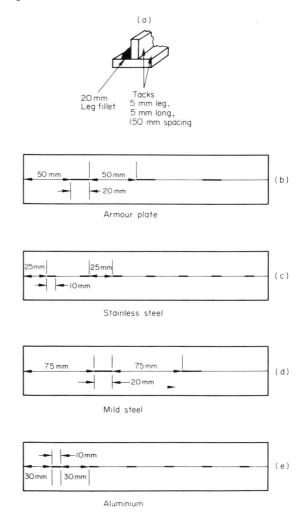

*Figure 2.29   Tacking — approximate tack lengths and pitch for light-gauge materials (butt joint)*

factors such as access, position in which the weld is to be made, process to be used and changes of material section must be assessed.

For square butt-preparations in materials from about 2 to 5 mm thick, as shown in figure 2.29, some general recommendations can be made. Tack welds on high-tensile material — figure 2.29b — should be of the same dimensions as the finished weld and of sufficient length to withstand stresses set up during the execution of the weld. Stainless steel (figure 2.29c) requires short close tacks, because it has a greater expansion coefficient than most other steels. It is also about five-times less efficient as a thermal conductor than mild steel; this means that excessive local heating, as in arc welding, could cause local parent-metal collapse, especially in very-light-gauge material. For this reason metal-arc welding is extremely difficult in material from about 0.8 to 2 mm thick and T.I.G. welding is most often used.

Mild steel presents less problems in tacking than most other materials, but it is still good practice to make tacks of the same dimensions as the finished weld and of adequate length, as shown in figure 2.29d. Aluminium has a high expansion-coefficient and low melting-point (660 °C). These factors mean that if a gap is allowed to develop in thin material, extensive local collapse is likely. Metal-arc welding is rarely used today for welding aluminium or its alloys, T.I.G. and M.I.G. (and for brazing, oxy-acetylene) being used instead. Tacks on thin aluminium should be short and executed as quickly as possible. The pitch, as shown in figure 2.29e will be determined partly by the heat built up in the area, thus a tack near the start of the weld, followed immediately by a tack at the start, may produce local collapse. Like copper and silver, aluminium is a very good thermal conductor and this characteristic should always be taken into account in welding.

**Presetting the Joint**

By trial and consequent determination of a satisfactory welding procedure and joint gap (figure 2.30), it is possible to produce a weldment consistently in close dimensional tolerance. The backstep sequence of deposition may be used along with an equal joint-gap as in figure 2.30b, or a continuous deposit along a tapered gap-joint as in figure 2.30a may be made. In this case, no root gap (figure 2.30c) will be left on completion of the weld.

The powerful contraction-forces developed by a cooling weld are shown diagrammatically in figure 2.31. A weld made around the perimeter of the plate, as in figure 2.31a will cause an increase in dimension *A* and a decrease in dimension *B* while a deposit on the inner circumference will have the reverse effect, as shown in figure 2.31b.

The restoration of a surface by metal-arc welding may be done in an intermittent mode or continuously as shown in figure 2.32, the latter method having the advantages of greater speed and lower incidence of weld craters at the plate edges. A neat finish can then be obtained by depositing the final weld around the plate perimeter.

A lap joint made on an assembly, as shown in figure 2.33, where a thin sheet overlays a comparatively thick section, requires particular attention to the electrode angle. If this is

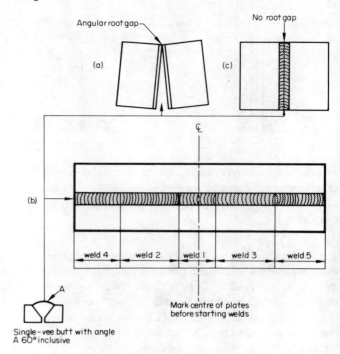

Figure 2.30    Distortion

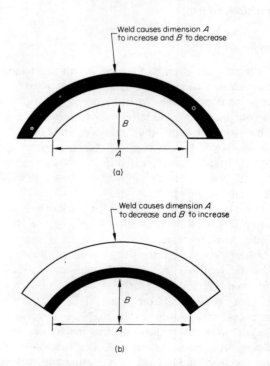

Figure 2.31    Weld-metal contraction

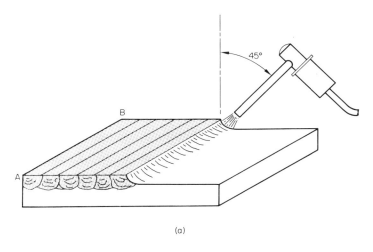

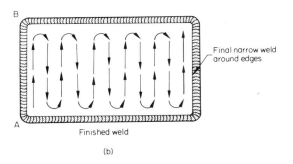

*Figure 2.32    Building up a surface by arc welding*

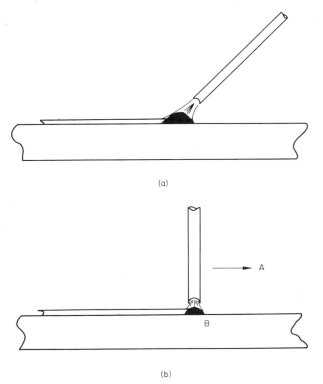

*Figure 2.33    Lap joint, thin sheet to thick plate — electrode angles; (a) incorrect — wrong electrode angle, long arc, (b) correct — correct electrode angle, short arc, lap joint thin sheet to thick plate*

too oblique, as in figure 2.33a, the top sheet will suffer from irregular melting and fusion along the weld edge of this side will be unsatisfactory. The electrode should be upright as in figure 2.33b and slightly disposed to the right (arrow A) so that preferential heat-input is directed towards the thick material. The edge of the weldpool, B, is then allowed to flow, rather than be directed to the adjacent thin sheet.

The usual method of producing a multi-fillet weld is to deposit the necessary number of welds of equal dimensions. Sometimes it is better to use the procedure shown in figure 2.34a, where weld 1 is made to the full fillet horizontal leg-dimension in one pass. This means that

1. a 'guide line' has been established for the successive runs

2. because a large electrode (and thus a greater heat-input) will be required, good root-fusion and pre-heating are achieved.

The weld in figure 2.34b will be difficult to make because of the acute joint-angle; there is a particular danger here of undercutting and slag trapping at A if the work is in the horizontal plane. Weld 2 should not be made until all slag inclusions have been removed from line B.

The reverse situation is seen in figure 2.35 where the wide joint-angle, although providing easy access, also lends itself to the deposition of welds having insufficient throat thickness (A—A). Undercut at B is also a risk, especially if weld 4 is made with excessive current or electrode weaving. The electrode angle must not be too greatly opposed to face C and the single-pass (weld 4) should not exceed 9.5 mm in leg length (in the h.v. position).

Multi-fillet welds in the overhead position are among the most difficult to produce owing to fatigue and the need to hold a short arc consistently — see figure 2.36. Each weld must be kept small in section — about as wide as the electrode diameter — no weaving should be used and as few breaks

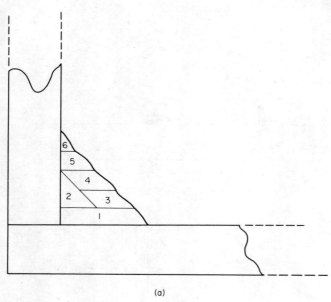

(a)

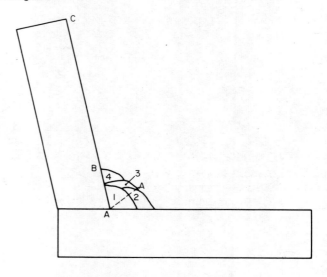

*Figure 2.35    Wide-angle h.v. multi-fillet sequence*

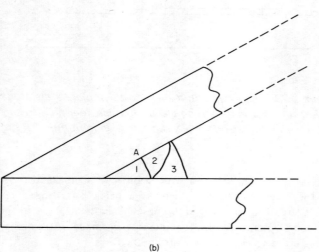

(b)

*Figure 2.34    Multi-pass    fillet    weld;    (a) h.v.    multi-fillet sequence, (b) acute angle h.v. multi-fillet sequence*

in deposition as possible are recommended. The sequence is different to that used in the h.v. position, the welds being made as in figure 2.36a. After weld 1 has been made, weld 2 is laid, preferably directly underneath it. The weld intersection point A then becomes a 'guideline' for the deposition of weld 3; similarly, the crowns of welds 3 and 4, at B and C, are useful in this way. If a poor fit-up condition exists as in figure 2.36b, weld 1 should be made as shown; thus

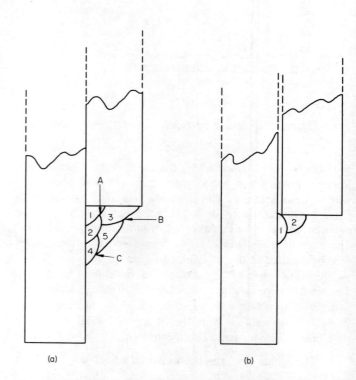

*Figure 2.36    (a) Sequence for overhead multi-fillet welding, (b) sequence for overhead multi-fillet welding with gap condition*

if it were made directly in the joint corner as in (a), the risks of undercutting, weld irregularity and slag inclusion, would be increased.

### Vertical Welding

Vertical welding needs a lot of skill and experience to perform satisfactorily, and a great deal of practice should be done in the f, h.v. and slightly inclined positions before attempting it. One of the limitations that has to be overcome is physical. In vertical welding it is a great advantage to be able to rest the weight of the body against a firm support and rest the elbow at a point near the work. This eliminates the fatigue that would quickly develop if you hold the welding cable, electrode and holder with an unsupported arm. Since welding requires a large number of small movements of controlled amplitude and precise direction to be made, the importance of support with regard to vertical welding can readily be understood.

The movements necessary for this work are illustrated in figure 2.37. There are two ways in which a vertical weld can be made in heavy-gauge material: in figure 2.37a a multi-sequence is used while in figure 2.37b a single pass is made. Both

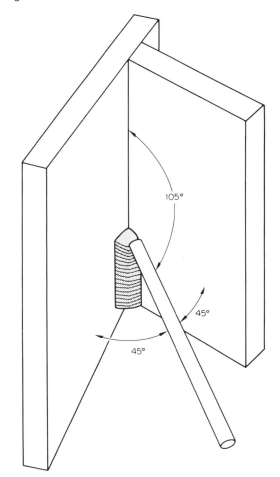

*Figure 2.38    Single-pass vertical fillet weld*

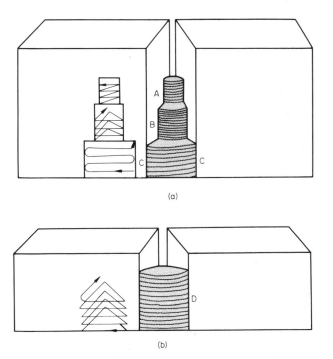

*Figure 2.37    (a) Three-pass vertical butt weld, (b) single-pass vertical butt weld*

methods have their advantages — mode (a) enables the welder to exercise greater control over distortion and also gives a slower rate of heat accumulation in the parent material; mode (b) has the advantage that greater and more uniform penetration can be achieved with less likelihood of slag inclusions.

Undercutting is one of the common faults apparent in vertical welds by inexperienced operators. In the case of a right-handed welder the more pronounced undercut will be on his right (C in figure 2.37a and D in figure 2.37b). The inwards (towards the body) movement possible is greater than the outwards one. In practice this means that movement towards the left is easier to execute, and so the rightward reach of the hand from the wrist may at times be insufficient.

A single butt or fillet weld as in figure 2.38 requires the use of correct electrode-angle, current and travel speed — the latter

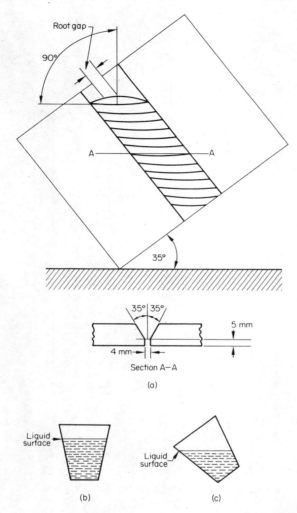

Section A—A

(a)

(b)                    (c)

*Figure 2.39    (a) Single-vee butt, inclined, (b) and (c) 'the liquid rule'*

*Figure 2.40    Single-pass vertical butt-weld on thick material*

only attainable by constant practice — while a single-vee butt-weld in the inclined (i) position (see figure 2.39a) underlines the necessity of working to the 'liquid rule', that is, if we tip a vessel containing liquid from the normal (upright, figure 2.39b) to the inclined (figure 2.39c), the surface of the liquid remains horizontal. This rule must be applied in the welding of vertical and inclined joints as shown in figure 2.40. The degree of inclination of the assembly in figure 2.39a may, of course, vary through almost $90°$ but this does not mean that any different procedure is required except for special reasons — such as the use of multiple welds, shown in figure 2.41, to build up a surface.

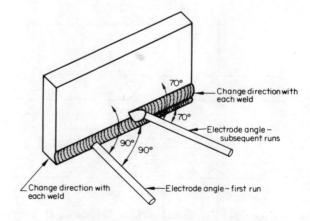

*Figure 2.41    Depositing overlay on a vertical surface*

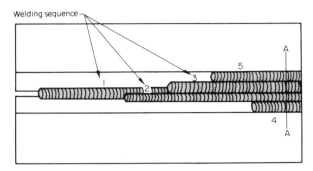

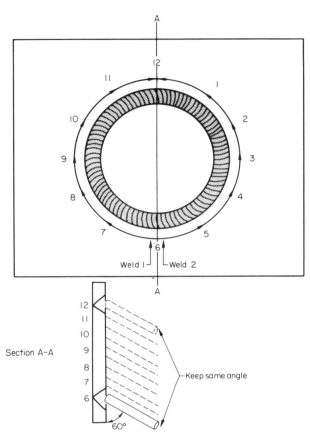

*Figure 2.42  Circumferential butt-weld in vertical position*

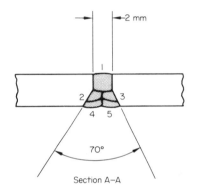

*Figure 2.43  Horizontal—vertical single-vee butt-weld*

A complete circumferential butt- or fillet-weld in the vertical position, as shown in figure 2.42, gives another application of the liquid rule (it is a common practice to use clockface numbers to denote positions in this case, and also when referring to pipe joints). In the horizontal—vertical position (figure 2.43) the liquid rule does not apply and multiple deposits in the case of large butt- or fillet-joints are necessary. Each weld bead should overlap the previous adjacent one by half its throat width.

The single-pass weld in figure 2.44 should not have leg lengths in excess of about 9.5 mm, otherwise difficulty will be met in maintaining equal profiles and undercutting may result along line A—A. Similarly undercutting can be caused in the flat (f) position (see figure 2.45) due to one, or a combination of, the following

1. excessively wide weave
2. current too high

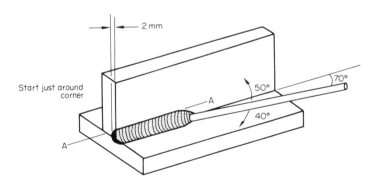

*Figure 2.44  Horizontal—vertical fillet-weld*

3. electrode angle incorrect
4. electrode size too small for weld width
5. work position (for example, welding down an incline)
6. arc too long
7. deflection of arc (d.c.) by local magnetic-fields
8. electrode flux-coating not truly concentric with core, causing arc bias to one side of weld
9. work position difficult to reach.

Figure 2.46 shows the weld correctly made, without undercut. Weld contour is as important as correct weld-size; failure of a joint under loading can often occur at the point of a weld defect as shown in figures 2.47a and b and figure 2.48.

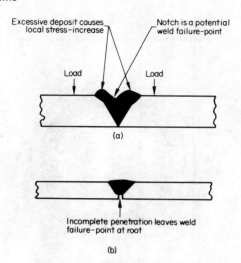

*Figure 2.47     Welding errors*

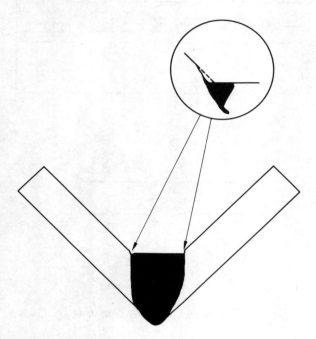

*Figure 2.45     Undercutting produced by excessive current, electrode-manipulation errors, difficulty of access to work area, arc too long; work position (that is, vertical or horizontal) may contribute greatly to this if work is executed by an inexperienced welder*

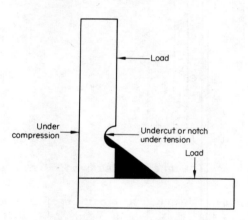

*Figure 2.48     Notch failure*

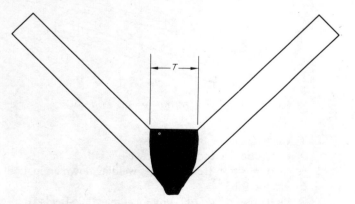

*Figure 2.46     Throat width determined by electrode manipulation, electrode size and type, arc voltage, travel speed and work position*

## Overhead Butt- and Fillet-joints

In heavy plate work, especially on site, or where the fabrication for some reason cannot be manipulated into a better position, large butt- or fillet-joints are sometimes necessary — see figure 2.49. Such joints must be made using the multiple-weld-bead method, as shown, because of the increased effects of gravity on a large volume of liquid metal (that is, below a certain volume the forces of liquid surface-tension can be utilised, as in overhead welding using small runs, to execute the work). Any attempt to exceed this threshold will not only result in the loss of the weldpool, but also in burns, possibly serious, to the welder.

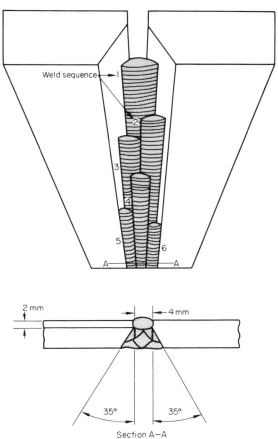

*Figure 2.49   Overhead single-vee butt-joint*

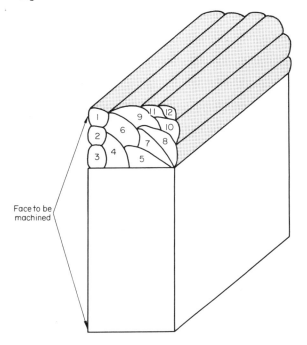

*Figure 2.50   Reclaiming section*

Machining is a costly item and any unnecessary work of this kind should be, and often can be, virtually eliminated through skill and forethought by the welder. An example is shown in figure 2.50; here a series of single beads is deposited on the side of the workpiece adjacent to the finished face; the addition of the remainder of the weld metal then follows. In this way the amount of weld metal projecting beyond the finished face is kept to a minimum. Afterwork may also be reduced or eliminated by using the method shown in figure 2.51 where clamps and machined packing-blocks are used to counter weld expansion- and contraction-forces.

### 'Stovepipe' Technique

The on-site welding of large-diameter pipes, using the metal-arc mode, employs a special technique. This is a method of

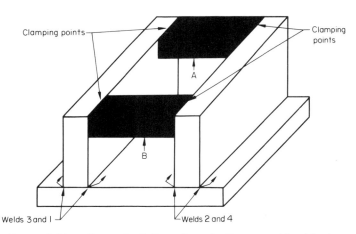

*Figure 2.51   Control of distortion; A, B, are packing blocks, length less than required finishing-width*

downward vertical welding based on the use of cellulosic deep-penetration electrodes. The first pass is made in the root, starting at the 12 o'clock position (figure 2.52) and proceeding to 6 o'clock. The procedure is then repeated of the other side of the pipe, followed by a 'hot pass' — a bead applied with a larger electrode at high current. The object of this is to wash

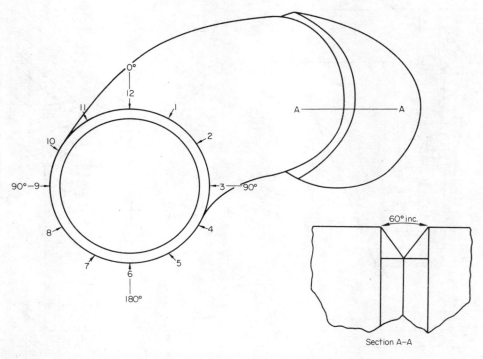

*Figure 2.52    The clock location-system*

out any trapped slag from the root run and also to increase the arc force and diameter to control and sustain a relatively large weldpool. The third or final pass is known as a capping bead and is applied using normal current. All passes begin at, or near, the 12 o'clock position and finish at 6 o'clock.

It is usual to employ teams of operators on pipeline construction, two welders to one joint, working simultaneously one at each side of the pipe. The welder in figure 2.53 is starting a run from the 12 o'clock position, using an engine-driven d.c. power source.

During the welding of a joint, the craters can be compared to a crucible holding molten metal (see figure 2.54). Here the analogy ends because a crucible is a leak-proof vessel holding a certain volume of liquid, whereas a weld crater can vary in size over a wide range, depending on current, arc force, position of the workpiece, parent-metal characteristics, arc length and voltage, etc. In addition, the crater is a travelling vessel that may leak from the bottom if the joint is not tight, or molten metal may be lost because the crater wall has broken down, as would happen in section A—A. Here, because the electrode (I) follows route E, the crater (H) will be breached. In section B—B, the movement of the crater (G) is arrested, because the electrode (J) follows route F.

Welding in the vertical position requires the maintenance of a slag front or dam (see figure 2.55) which is, in effect, one wall of the crater or crucible. If the slag dam is lost by wrong electrode-manipulation, the dam is destroyed and the weld pours out of the crater. In figure 2.56 the angle of the electrode used in vertical welding may be compared to that in figure 2.57 where a weld is being made in the flat position.

**Plug Welding**

This method of local fusion of two or more metal sheets can be accomplished by using one of a number of techniques, for example, metal-arc, T.I.G., M.I.G., submerged-arc, carbon-arc, etc. Since metal-arc welding produces a slag coating on the deposit, the process is more limited in application than, say, M.I.G. or T.I.G. Whatever welding method is used, however, there are certain common conditions to be observed, see figure 2.58. For example, the thinnest plate should always be uppermost, otherwise burn-through (figure 2.58a) will occur. The correct assembly is shown in figure 2.58b. For metal-arc welding thin to thick material (as in figure 2.58c) a hole of about 5 mm diameter is an advantage, but if the top sheet is,

Figure 2.53  Pipeline welding — 'stovepipe' welding line pipe using G.K.N. Lincoln Electric Ltd's mobile welding-units and electrodes (courtesy of the Lincoln Electric Co. Ltd)

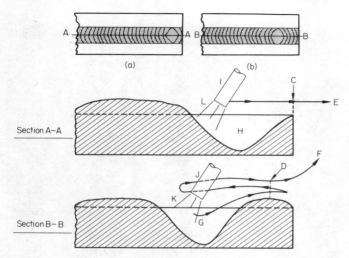

*Figure 2.54    Manipulation technique: C, D, break arc points, E, F, electrode routes, G, H, craters, I, J, electrodes, K, L, arcs*

*Figure 2.56    Vertical welding (courtesy of B.O.C. Ltd)*

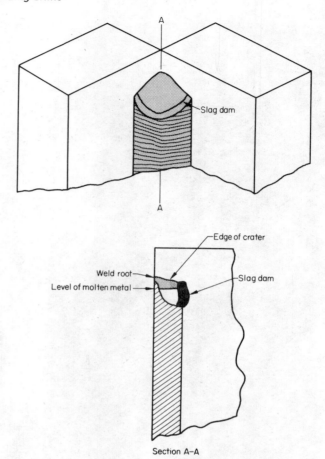

*Figure 2.55    Vertical fillet-weld*

say, 12 mm thick, then a hole of about 15 mm diameter, or one with a wide taper as in figure 2.58d, is required to provide room for slag manipulation. Some other preparations are shown in figure 2.59.

## FURTHER STUDY

Joseph William Giachino and W. Weeks, *Welding Technology*, Technical Press, Oxford, 1968

Alfred Haas, *Basic Industrial Electronics*, Iliffe, London, 1964

Mel M. Schwartz,Mel M. Schwartz, *Modern Metal Joining Techniques*, Wiley, New York, 1969

Eric Norman Simons, *Dictionary of Ferrous Metals*, Muller, London, 1972

R. E. Smallman (Ed.), *Modern Physical Metallurgy*, Butterworth, London, 1970

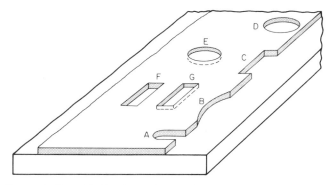

*Figure 2.59 Weld through plate and edge preparations; A — radial slot, B — elliptical slot, C — square slot, D — drilled or punched hole, E — tapered hole, F — square-edged slot, G — tapered slot*

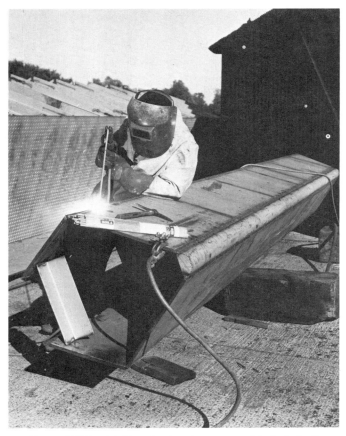

*Figure 2.57 Welding in situ (courtesy of B.O.C. Ltd)*

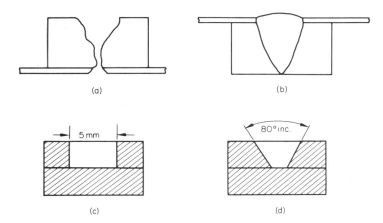

*Figure 2.58 Plug welding*

K. G. Richards, *Joint Preparations for Fusion Welding of Steel*, Welding Institute, Cambridge, 1966

P. F. Woods, *Production Welding*, McGraw-Hill, Maidenhead, 1971

Engineering Industry Training Board, *Argon-arc Welding*, Booklet 3/15, Watford, 1970

Engineering Industry Training Board, *$CO_2$ Consumable-electrode Arc-welding*, Booklet 3/13, Watford, 1973

Engineering Industry Training Board, *Manual Electric-arc Welding*, Booklet 3/14, Watford, 1973

Industrial Safety Manufacturers Association, *Reference Book of Protective Equipment*, London, 1973

Welding Institute, *Control of Distortion in Welded Fabrications*, Cambridge, 1963

Wiggin Nickel Alloys Co., *Welding, Soldering and Brazing*, Publication 3367, London, 1968

H.M.S.O. Booklet 13, *Ionization Radiations, Precautions for Industrial Users*

H.M.S.O. Safety Pamphlet 18

BS 1500: Fusion welded pressure vessels for general purposes: Part 3:1965 Aluminium

BS 2633:1973 Class 1 arc welding of ferritic steel pipe work for carrying fluids

BS 2971:1961 Class 2 metal-arc welding of steel pipelines and pipe assemblies for carrying fluids

BS 4570: Fusion welding of steel castings: Part 1:1970 Production, rectification and repair

**Free issue of periodicals of interest to engineers**

*$CO_2$ Welding News*, Distillers Company (Carbon Dioxide) Ltd, Reigate, Surrey

*Flare,* West Midlands Gas, Solihull, Warks.

*Hobart Weldworld,* Hobart Equipment Suppliers, Leeds

*The Nickel Bulletin,* International Nickel Co., Millbank, London SW1

*S.I.F. News,* Sifbronze Ltd, Stowmarket, Suffolk

*Stabilizer,* Lincoln Electric Co. Ltd, Welwyn Garden City, Herts.

*Stubs Welding Chronicle,* Stubs Welding Ltd, Warrington, Lancs.

*Svetsaren* (English edition), E.S.A.B. Ltd, Gillingham, Kent

*The Welder,* B.O.C.-Murex Ltd, Waltham Cross, Herts.

**Trade journals free to production engineers, welding engineers and managers**

*Compressed Air and Hydraulic Engineering,* Ingersoll-Rand Co. Ltd, Knightsbridge, London SW1

*Design Engineering,* Morgan-Grampian (Publishers) Ltd, Calderwood Street London SE18

*The Engineer,* Morgan-Grampian (Publishers) Ltd, Calderwood Street, London SE18

*Factory and Industrial Equipment,* New Media Ltd, Argylle Street, London W1

*Iron Age Metalworking International,* Jermyn Street, London SW1

*Metalworking Production,* Morgan-Grampian (Publishers) Ltd, Calderwood Street, London SE18

*Sub-Assembly Components Fastening,* Commercial Exhibitions & Publications Ltd, Hough Street, London SE18

# 3. OXY-ACETYLENE TECHNIQUES

## OXY-ACETYLENE EQUIPMENT

Although the oxy-acetylene processes have been superseded in great measure by modern techniques like M.I.G. welding and plasma-arc cutting, they are still, because of their versatility, used in some industries; car manufacturers, for instance, still use the process and almost every garage and small workshop has its oxy-acetylene welding and cutting equipment.

The basic equipment consists of

1. a welding torch that is safe, light and efficient, with easily adjustable controls and the facility for quick change of nozzles to give a wide range of heating power

2. cylinder regulators with two stages of pressure — (a) the cylinder pressure and (b) the outlet or working pressure

3. protective goggles of a range of shades complying with BS 679

4. filler rods of various metals and alloys from 2 to 5 mm diameter and supplied, for convenience of handling, in lengths of 1 m

5. flexible high-pressure hoses, gas economiser and back-pressure valves.

### The Low-pressure System

There are two systems employed in the use of oxygen and acetylene, these are the *low-pressure* and the *high-* or *equal-pressure* methods. Both produce the same flame-characteristics, the differences otherwise being in the method of supplying acetylene and the construction of the welding blowpipe.

If acetylene is produced using a generator in which gas is made from calcium carbide and water, low-pressure torches must be used, but for compressed dissolved acetylene, supplied in special cylinders, high-pressure blowpipes are necessary. *High-pressure (or equal-pressure) blowpipes must never be used with a low-pressure system*, to do so would be extremely dangerous.

When acetylene is produced in a generator the maximum pressure must not exceed 1550 kg/m$^2$ (15 kN/m$^2$). To obtain correct conditions for combustion, the gases must be pre-mixed in the blow-pipe. This is done by using an injector which injects a high-pressure stream of oxygen into a surrounding low-pressure stream of acetylene. The low-velocity stream of acetylene is thereby sucked into the oxygen stream which enters the mixing chamber before issuing from the tip or nozzle. At this point, both gases are of approximately equal proportions and velocity. The correct injector and tip relationship must be observed, so with the low-pressure system the whole tip-and-shank assembly must be changed when altering the blowpipe's output heating-power. The low-pressure blowpipe requires about 1.14 volumes of oxygen to 1 of acetylene.

### The High- or Equal-pressure System

This has almost entirely superseded the older low-pressure mode. With high-pressure equipment, greater mobility and usage range are possible, as well as an increased margin of equipment safety. The high-pressure blowpipe is of lighter and simpler construction and backfires, with their inherent risk, are far less frequent.

No injector is necessary since the gases are piped to the torch at equal pressures. To change the heating power of the blowpipe only the tip need be replaced together with an adjustment of gas-supply pressures.

Each welding tip at neutral-flame setting provides an average gas velocity of about 90 m/s. Acetylene and oxygen are provided in cylinders of varying sizes. Most acetylene cylinders have a capacity of about 6250 litres. Oxygen can be provided in gaseous form in cylinders or from a bulk liquid-supply tank via an evaporator. Not more than 20 per cent of the contents of an acetylene cylinder should be drawn off per hour, otherwise the solvent liquid (acetone) will be vapourised. Manifold systems, using interconnected cylinders of dissolved acetylene and a bulk supply of oxygen are subject to Home Office regulations as regards operation and storage. Acetylene tanks must be stored in well-ventilated properly constructed buildings, situated away from the main buildings of a factory if possible. If the oxygen supply is provided by cylinders, these must also be stored in a separate building well away from the acetylene store.

Acetylene is distributed through steel pipes; copper must never be used because of the danger of producing the violently explosive compound copper acetylide $(Cu_2C_2)$; silver also forms an explosive compound with acetylene. Oxygen-distribution pipelines are made of copper with bronze or stainless-steel fittings.

Acetylene cylinders are charged with gas at a pressure of $14 \times 10^4$ kg/m$^2$ (1380 kN/m$^2$). Free acetylene, such as acetylene in a pipeline system, must never exceed a pressure of $1.05 \times 10^4$ kg/m$^2$ (103 kN/m$^2$), otherwise it may detonate. The gas is only safely stored at high pressure if it is pure, dry and dissolved in acetone in a cylinder. Acetone is absorbed in kapok, carbon in the form of charcoal, or asbestos. Acetone, $(CH_3)_2CO$, is a pungent-smelling liquid which boils at 56 °C, and is slightly toxic. It is inflammable and absorbs 25 times its own volume of acetylene at atmospheric pressure. When the pressure is increased to $14 \times 10^4$ kg/m$^2$ (1380 kN/m$^2$) it absorbs 370 times its own volume of acetylene.

## Gas-pressure Regulators

There are two types of regulator: single stage, and two stage.

In the single-stage regulator the gas from the cylinder enters the low-pressure chamber by adjusting the pressure-control screw, forcing the diaphragm back against the spring pressure. The strength of the spring is calculated so that it contains the pressure sufficiently to allow the quantity of gas required at the blowpipe to pass into the pipeline.

### The Two-stage Regulator

This uses two diaphragms and two control valves so that the pressure adjustment from cylinder to blowpipe is made in two separate stages.

Because of its great versatility and low capital cost compared with other welding processes, oxy-acetylene equipment is still good value for money and, indeed, in many instances it is still commonly used. Some major uses are in small-bore-pipe fabrication, sheet-metal manufacture, jewellery-making, and the motor industry.

Figure 3.1 shows a light-weight modern welding-torch that, besides being extremely efficient, has excellent ergonomic design features. The control valves are situated close together so that it is easy for the welder to make thumb adjustment to the flame during welding. The base of the torch is of rectangular section with fluted side-panels providing a sure grip for a gloved hand. By changing the tip assembly it is possible to use the torch for flamecutting, as shown in figure 3.2.

For extremely fine work such as brazing, welding and soldering of jewellery and dental components, etc., a micro-torch (figure 3.3) is ideal. The welding tips are fitted with synthetic drilled rubies, the smallest tip has an orifice diameter of only 0.1 mm (0.004 in.) and the largest, 0.74 mm (0.029 in.). The welding kit, shown in figure 3.4, can be used with a number of fuel gases and oxygen, for example, acetylene, propane, hydrogen, coal gas and natural gas.

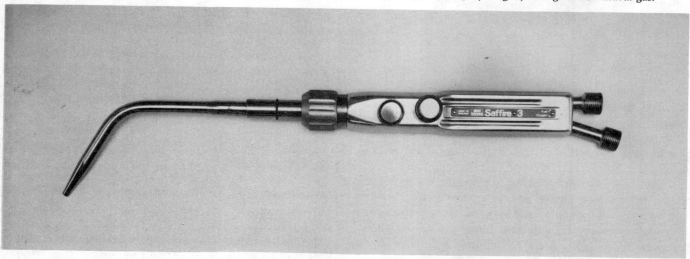

Figure 3.1    Light-weight torch (courtesy of B.O.C. Ltd)

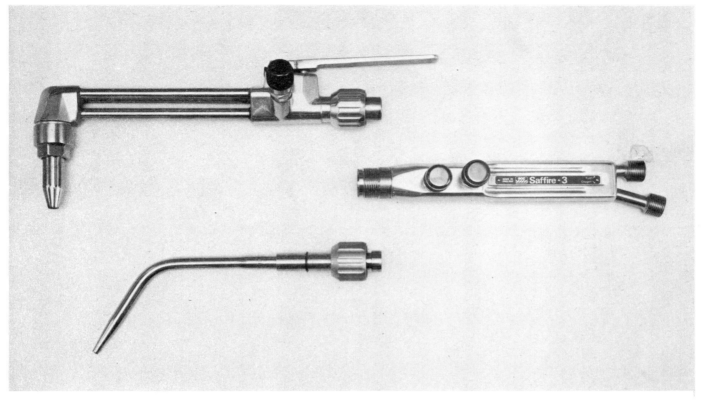

Figure 3.2    Combined cutting and welding torch (courtesy of B.O.C. Ltd)

## Oxygen and Acetylene

Oxygen is a diatomic gas (chemical symbol $O_2$) produced commercially from the fractional distillation of liquid air. The gas is extremely reactive and readily forms compounds with many other elements. Only about 23 per cent of the air is oxygen, the bulk of the remainder being nitrogen (75.5 per cent) with traces of argon (1.3 per cent) and even smaller amounts of carbon dioxide, krypton, xenon, neon and helium. There are variable trace amounts of other gases, including hydrogen and ozone, and, in an industrial atmosphere, sulphur dioxide and oxides of nitrogen.

Separating air into its component gases is done in three stages: compression, cooling and evaporation. Each liquid gas has its own boiling point, for example, nitrogen boils off at $-196\,^\circ$C, oxygen at $-183\,^\circ$C.

Acetylene, $C_2H_2$, is a compound of carbon and hydrogen and is produced in bulk by the reaction of water with calcium carbide. Most of the calcium carbide used in the manufacture of acetylene in the United Kingdom is derived from limestone,

which is usually processed in its country of origin; cheap hydro-electric power makes the production of high-quality carbide possible at Tyssedal near Odda in Norway.

Figure 3.5 shows the basic equipment for oxy-acetylene welding. Acetylene is always provided in a maroon-painted cylinder, oxygen in a blacked-painted cylinder.

Acetylene cylinders have a left-hand-thread outlet valve, while oxygen cylinders have a right-hand-thread outlet valve. Each cylinder is fitted with a two-stage pressure regulator. *It is most important that pressure regulators are used only for the gas for which they are specified.*

Let us suppose that the apparatus illustrated has not been assembled; how do we go about fitting it up correctly and *safely*?

## Assembly of Equipment

First we must make sure that the cylinders are *secured* in an appropriate trolley or stand, and in each case fastened by a

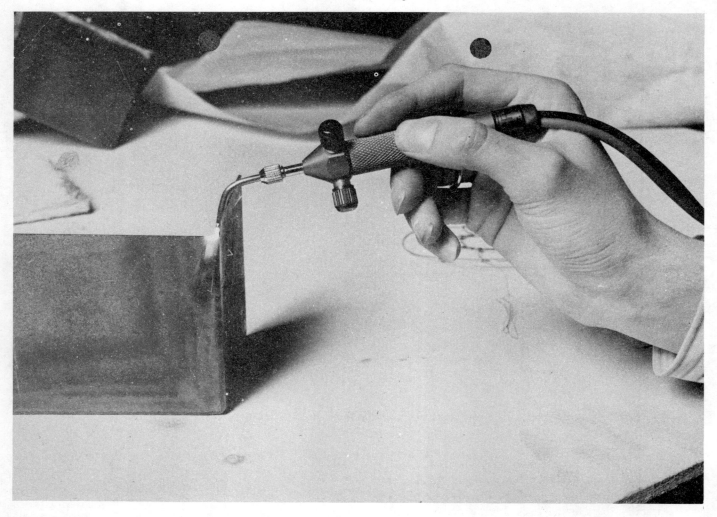

*Figure 3.3     Miniature torch (courtesy of C. S. Milne and Co. Ltd)*

chain or locking bar. Cylinders must not be used when they are lying on the floor or located near a heat source such as an open fire, radiator or boiler.

Next we must screw the appropriate pressure-regulator into each cylinder, but before we do this, the cylinder valves should each be *slowly* and *briefly* opened and shut to blow out any dust from the valve seatings. (*Note* that no lubricant of any kind should be used on any part of oxy-acetylene apparatus — oil and grease are especially dangerous in the presence of oxygen.) When the regulators have been fitted, the hoses (blue for oxygen, maroon for acetylene) can be connected to the second stages of the pressure regulators. At the blowpipe end of the hoses check valves should be fitted (A—A in figure 3.5)

to protect the hoses from flash-back which could generate sufficient pressure to burst the pipes.

**Lighting-up Procedure**

1. *Slowly* open the cylinder valves (sudden opening may damage the pressure regulators).

2. Adjust the regulator screws to obtain the correct second-stage pressure.

3. Open each blowpipe valve in turn to purge the system of air.

4. Check that the working pressure (second-stage regulator) is correct.

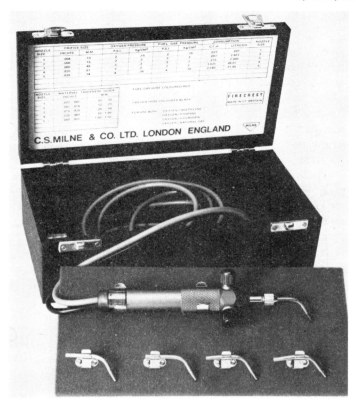

*Figure 3.4    Miniature-torch kit (courtesy of C. S. Milne and Co. Ltd)*

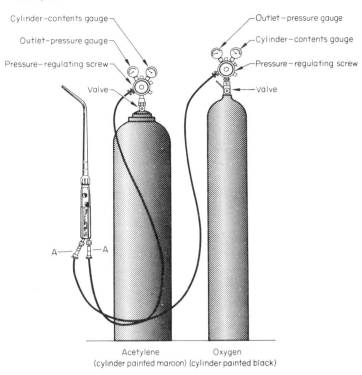

*Figure 3.5    Cylinder details*

5. Close both blowpipe valves.

6. Open the acetylene valve gently ensuring that a sufficient flow of acetylene is provided.

7. Light up; if the flame burns away from the tip or 'lifts off', adjust the valve until the flame touches the tip (thus it is always safer to start with an excess flow).

8. Open the oxygen valve slowly until the inner flame is sharply defined — this is the neutral flame.

**Shutting-down Procedure**

1. Close the acetylene valve on the torch.
2. Close the oxygen valve on the torch.
3. Close cylinder valves.
4. Open and close torch valves to relieve pressure in the hoses — oxygen first, then acetylene. Place torch on an appropriate stand — do not leave it lying on the workbench or floor.

The same general procedure and safety precautions also apply to cutting- and heating-blowpipes.

## SAFETY

*Before using any oxy-acetylene apparatus the safety rules should be observed and understood.*

*Always follow this procedure:* Set the regulators to the recommended working-pressure. Keep the blowpipe nozzle away from any source of ignition (pilot light, smouldering tow, etc.) until the fuel gas is flowing freely from the nozzle. The use of a spark lighter is recommended for lighting blowpipes; if the blowpipe flashes back on lighting up it is because

(i) the regulators are not set to the correct pressure, or

(ii) a light has been applied before the flow of fuel gas is properly established.

If the flame snaps out when the blowpipe is in use, it is because

(i) the regulator pressure and/or gas flow are incorrect, either too high or too low

(ii) the nozzle has been obstructed

(iii) the nozzle is held too close to the work

(iv) the nozzle has become overheated.

*When this happens:* Completely shut both blowpipe valves, check regulator setting, cylinder pressures, and relight in accordance with procedure given above. In the case of (iv), plunge nozzle and blowpipe head into cold water. Make sure that the nozzle is tight before re-lighting the blowpipe.

At all times see that cylinders are protected from rusting, heat and corrosive conditions. Do not lay cylinders on wet soil.

### See that cylinders are stored properly

If cylinders are exposed to heat the cylinder walls may be weakened, and at the same time the pressure of the gas content will increase and dangerous conditions may result. Therefore store them well away from sources of heat such as furnaces, stoves, boilers and radiators, and away from combustible material and blowpipe flames. Oil or grease will ignite violently in the presence of oxygen, and if the oxygen is under pressure an explosion may result. Cylinders and fittings should be kept away from all sources of contamination such as oil barrels, overhead shafting, cranes or belts.

Do not smoke, wear oily or greasy clothes or have any exposed light or fire in any place where compressed gases are stored. Do not handle oxygen cylinders, valves or any fittings with greasy hands, gloves or rags.

Oxygen has no smell, and while it does not burn, it supports and accelerates combustion. Ordinary clothing and more inflammable materials such as oil can become ignited by a spark, and will burn fiercely in oxygen or where the atmosphere has become enriched with oxygen. In a confined space a small amount of acetylene or oxygen may create a dangerous condition that will cause fire from a spark or naked light.

Take care to avoid leakages; test with soapy water and a brush or a 5 per cent solution of Teepol — *never test with a naked flame.*

Acetylene can form explosive compounds when in contact with certain metals or alloys, in particular those of copper and silver. Joint fittings or piping made of copper should on no account be used, and acetylene should never be allowed to come into contact with copper or any alloy containing more than 70 per cent copper.

Use maroon or red hose for acetylene and other combustible gases, and blue for oxygen — be careful to see that they are never interchanged. Use hoses of equal length and do not coil any surplus hose around regulators or cylinders.

Do not use odd bits of tubing and remember that copper or high-copper-content alloy must never be used in acetylene

hose or other parts in contact with acetylene; use a proper adaptor.

Be sure to observe carefully the makers' instructions for lighting and using blowpipes. Do not use pressures in excess of those recommended or use heavy-duty high-delivery regulators where only low pressures are required for operating blowpipes. Never attempt to light a blowpipe until sufficient time has elapsed, after opening the blowpipe acetylene-valve, for the gas in the hose to normalise at the correct working-pressure and for all air to be displaced from the hose.

The safety precautions (listed by the manufacturer) to be taken in the event of a cylinder becoming heated up due to a backfire or other incident, should always be followed. Failure to carry out these instructions and precautions properly may cause the cylinder to heat up internally and burst.

Ample means of thorough ventilation should constantly be maintained in welding shops, and special provision should be made for ventilating confined spaces in which oxy-acetylene apparatus is used.

Fire extinguishers and sand should be readily accessible, and water should be used to flood the floor where no other protection is possible. When working near wooden constructional work, the wood should be carefully protected to prevent any contact with the flame or hot metal. Where there is danger of smouldering fires, special precautions should be taken to prevent trouble after work is finished.

### Working on Tanks or Vessels

Do not weld or cut tanks or vessels that may have contained petrol, oil, spirits, paint, or any inflammable or explosive material, without first making sure that the vessel contains no trace of the substance or any explosive vapours, and that the vessel has been treated to make them safe.

### Working on Painted Surfaces

Wear an approved respirator when flame-cleaning in confined spaces, paint-burning, cutting or welding on painted or galvanised plate for an extended time (or even for a short time in a confined space).

All oxy-acetylene welders must use goggles to protect their eyes from sparks and to prevent eye strain. The most efficient form of lenses are those which conform to BS 679:1959 Filters for use during welding and similar industrial operations. Goggles with inflammable lenses or frames should not be used. Never allow cylinders, their colour, valve, threads or markings, to be altered or tampered with in any way. Users should refuse to accept delivery of any gas cylinder whose contents are not clearly indicated by labelling, marking, or colour. Never attempt to mix gases in a cylinder or try to fill one cylinder from another. *Call each gas by its proper name; do not refer to oxygen as 'air', or to acetylene as 'gas'.*

## GASES

### Oxy-acetylene

The oxy-acetylene flame is dominant in the chemical-flame group of processes. Other combinations of gases cannot equal its unique characteristics, for example, it is the only one that produces a truly neutral flame condition, others tend to be oxidising or carburising and none can equal its high temperature, heat input and flame-propagation rate.

In the combustion of oxygen and acetylene, the gases are mixed in the torch in about equal volume (in the equal-pressure or high-pressure system), the rest of the oxygen required to support the flame being drawn from the atmosphere. The ratios are: one volume of acetylene plus one volume of oxygen, supplied from the cylinders, and one-and-a-half volumes of oxygen from the atmosphere. Because all of the oxygen required does not come from the cylinder, the following primary reaction occurs in the inner cone of the flame adjacent to the welding tip

$$2C_2H_2 + 2O_2 \longrightarrow 4CO + 2H_2$$

that is, two volumes of oxygen plus two volumes of acetylene burn to produce four volumes of carbon monoxide plus two volumes of hydrogen. This initial exothermic reaction is responsible for the brilliant blue intense inner cone, while the secondary combustion products, carbon monoxide and hydrogen, consume atmospheric oxygen to provide the secondary flame according to the following reactions

$$4CO + 2O_2 \longrightarrow 4CO_2$$
$$2H_2 + O_2 \longrightarrow 2H_2O$$

As with most flames burning in air, water vapour and carbon dioxide are the end products of combustion.

The most used and useful part of the oxy-acetylene flame is the primary or inner cone, at the tip of which the highest temperature is attained (3200–3300 °C, depending on gas purity and mixing). The velocity of the gas is lowest when passing through the tip bore, because the wall provides resistance to the flow. Thus the inner cone is short and the greatest gas-velocity is reached in the centre of the tip bore

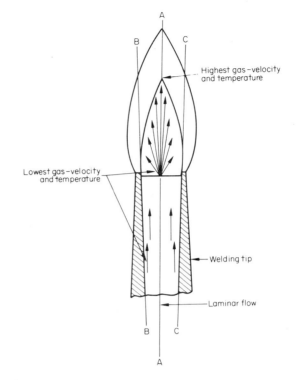

*Figure 3.6   Flame characteristics; A–A, gas in bore centre has highest velocity; B–B, C–C, bore-surface friction, gas at lowest velocity*

(that is, furthest from the walls), and so the flame is longest at this point – see figure 3.6. The cone shape thus depends on the characteristics of the final bore-diameter of the tip (regularity of diameter, smoothness, etc.) and those of the preceding supply-pipe (the ratio between the neck and tip diameters, etc.). A small orifice or tip provides a sharp inner-cone while progressively larger tips create a rounded type of cone – see figure 3.7.

Manufacturers of equipment provide tables giving gas flow-rates, pressures and thermal ratings for each tip diameter, and these can be correlated with the thickness of metal to be welded. The recommended gas-pressures should be observed, for example, excessive pressure may displace molten metal from the weldpool and generally increase the difficulty of welding, while insufficient pressure may cause the flame to burn downstream or 'pop back' into the torch mixer-unit. Each tip, in terms of heat input, overlaps the next and this latitude can be exploited by the skilled welder, who may, for instance, quite successfully weld 1 and 3 mm thick material by adjusting his flame parameters.

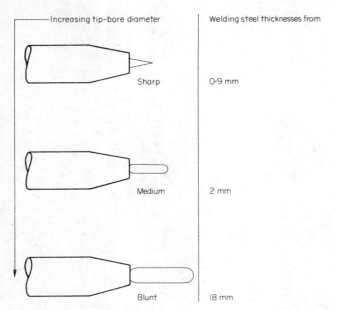

| | Welding steel thicknesses from |
|---|---|
| Sharp | 0·9 mm |
| Medium | 2 mm |
| Blunt | 18 mm |

*Figure 3.7   Tip sizes*

For welding steel the neutral-flame setting is mostly used, but a slightly carburising flame may sometimes be employed. Here the addition of carbon to the molten pool lowers the melt temperature slightly, giving a small increase in welding speed, but for most welding applications the neutral flame is still the most satisfactory.

*Blowback*

If the issuing-gas velocity from the torch is insufficient to maintain the flame (that is, it is less than the combustion velocity) the flame will burn downstream ('blowback'). Conversely, should the issuing velocity exceed the combustion velocity, the flame will lift off the tip. The point at which the flame base impinges on the tip is called the boundary layer and it is here that flame equilibrium is maintained by the steady laminar flow of unburnt gases. The rate of change in mixture velocity in the boundary layer is called the boundary velocity-gradient and this is determined by the conditions of flow, which may be laminar or turbulent. Blowback may also occur if the tip is not tightly screwed into the torch body, or if it has been damaged or the bore obstructed in any way. In the event of blowback, the torch should be extinguished and not re-lit until the cause has been found and the trouble rectified.

The great versatility of the oxy-acetylene flame may be judged from the variety of materials that can be welded; these include iron and its alloys, copper and its alloys, aluminium and its alloys, platinum, silver, gold, cast iron, lead, zinc and glass.

**Air—Acetylene**

This flame only finds application in the welding of lead, light-brazing and soft-soldering.

**Oxy-hydrogen**

This can be used for welding aluminium and some alloys, particularly in thinner gauges. It may also be used for lead welding (often wrongly referred to as 'lead burning') in material from about 5 to 12 mm thick. Magnesium and its alloys are weldable with oxy-hydrogen but, like aluminium, most work is now done with T.I.G. (G.S.T.A.).

**Hydrocarbon Gases**

Hydrocarbon gases such as butane, propane and coal gas are not suitable for welding ferrous material owing to their oxidising character. These fuel gases have relatively low flame-propagation rates, with the exception of propane and coal gas (which contains about 50 per cent hydrogen). Methane (which constitutes about 94 per cent of 'natural gas'), butane, and propane, with small amounts of other gases such as ethane and propadiene (allene), are sometimes used, with oxygen, for flamecutting but their chief use is in industrial heating-applications. Propane is widely used for this purpose but mixtures of hydrocarbon gases are marketed commercially under such names as M.A.P.P., A.P.A.C.H.E., and 'Calor'.

## OXY-ACETYLENE WELDING TECHNIQUES

The *leftward* welding-method is most commonly used, the *rightward* mode being restricted to heavy-gauge-material fabrication. In leftward welding (figures 3.8 and 3.9) the torch is held steadily in the right hand while filler is added in rhythmic sequence. Joint included angles should be 80 to 90° for thick material, the upper limit for a closed square butt being plate of about 5 mm thickness.

When starting to weld, it is important to realise that fusion of the parent material must occur before adding filler metal. The general procedure is as shown in figure 3.10. Premature melting of the filler wire, and its addition to the work before a melt area has been established, will result in lack of fusion. To begin with, only a small portion of the filler wire should be deposited and allowed to wet the surface, forming a small hemisphere or pool.

55

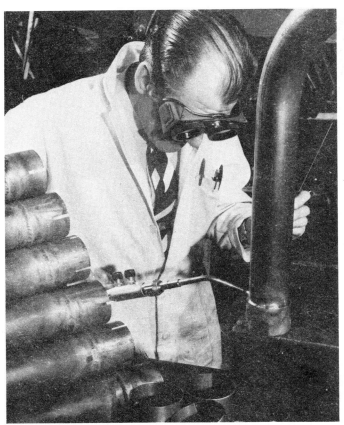

Figure 3.8    Gas-welding technique (courtesy of B.O.C. Ltd)

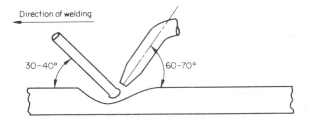

Figure 3.9    Leftward welding

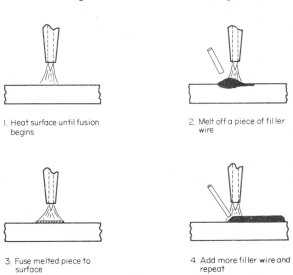

1. Heat surface until fusion begins

2. Melt off a piece of filler wire

3. Fuse melted piece to surface

4. Add more filler wire and repeat

Figure 3.10    Oxy-acetylene welding technique

Table 3.1    Conditions for welding in relation to material thickness using high-pressure blowpipe (data courtesy of B.O.C. Ltd)

| Mild-steel thickness | | | Nozzle size | Operating pressures | | | | Gas consumptions | | | |
|---|---|---|---|---|---|---|---|---|---|---|---|
| | | | | Acetylene | | Oxygen | | Acetylene | | Oxygen | |
| mm | in. | S.W.G. | | $N/m^2 \times 10^5$ | lbf/in.$^2$ | $N/m^2 \times 10^5$ | lbf/in.$^2$ | l/h | ft$^3$/h | l/h | ft$^3$/h |
| 0.9 | — | 20 | 1 | 0.14 | 2 | 0.14 | 2 | 28 | 1 | 28 | 1 |
| 1.2 | — | 18 | 2 | 0.14 | 2 | 0.14 | 2 | 57 | 2 | 57 | 2 |
| 2 | — | 14 | 3 | 0.14 | 2 | 0.14 | 2 | 86 | 3 | 86 | 3 |
| 2.6 | — | 12 | 5 | 0.14 | 2 | 0.14 | 2 | 140 | 5 | 140 | 5 |
| 3.2 | 1/8 | 10 | 7 | 0.14 | 2 | 0.14 | 2 | 200 | 7 | 200 | 7 |
| 4 | 5/32 | 8 | 10 | 0.21 | 3 | 0.21 | 3 | 280 | 10 | 280 | 10 |
| 5 | 3/16 | 6 | 13 | 0.28 | 4 | 0.28 | 4 | 370 | 13 | 370 | 13 |
| 6.5 | 1/4 | 3 | 18 | 0.28 | 4 | 0.28 | 4 | 520 | 18 | 520 | 18 |
| 8.2 | 5/16 | 0 | 25 | 0.42 | 6 | 0.42 | 6 | 710 | 25 | 710 | 25 |
| 10 | 3/8 | 4/0 | 35 | 0.63 | 9 | 0.63 | 9 | 1000 | 35 | 1000 | 35 |
| 13 | 1/2 | 7/0 | 45 | 0.35 | 5 | 0.35 | 5 | 1300 | 45 | 1300 | 45 |
| 19 | 3/4 | — | 55 | 0.43 | 6 | 0.43 | 6 | 1600 | 55 | 1600 | 55 |
| 25 | 1 | — | 70 | 0.49 | 7 | 0.49 | 7 | 2000 | 70 | 2000 | 70 |
| 25+ | 1+ | — | 90 | 0.63 | 9 | 0.63 | 9 | 2500 | 90 | 2500 | 90 |

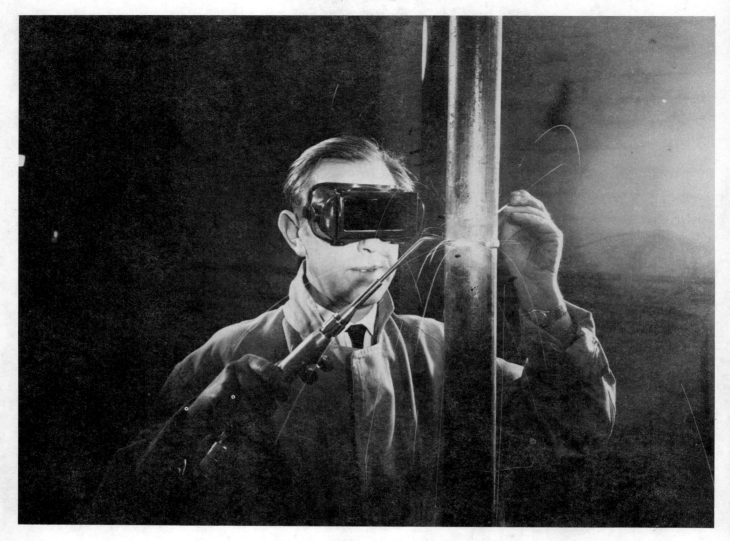

*Figure 3.11    Pipe welding (courtesy of B.O.C. Ltd)*

Using the rightward technique welding commences at the
left-hand end of a joint and proceeds towards the right.
Heavy-gauge material can be welded up to 15 mm thick, and in
the vertical position, using two operators, up to 25 mm. The
relative positions of filler rod and torch for rightward welding
can be seen in figures 3.11 and 3.12. Vertical welding requires
greater inclusive joint angles, see figure 3.13.

Edge-welding preparations and relative welding-speeds between leftward and rightward welding are shown in figure-
3.14.

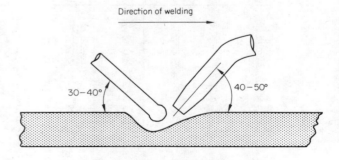

*Figure 3.12    Rightward welding*

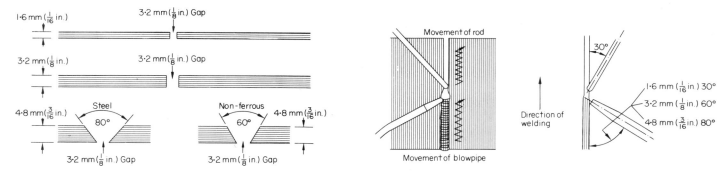

Figure 3.13 Vertical welding, joint preparation and technique (data courtesy of B.O.C. Ltd)

| Thickness of metal | Diameter of welding rod | Edge preparation | | Speed mm/min | Thickness of metal |
|---|---|---|---|---|---|
| Less than 0·9 mm (20 S.W.G.) | 1·2–1·6 mm ($\frac{3}{64}-\frac{1}{16}$ in.) | | Leftward welding | 127–152 | 0·8 mm ($\frac{1}{32}$ in.) |
| | | | | 100–127 | 1·6 mm ($\frac{1}{16}$ in.) |
| 0·9–3 mm (20 S.W.G. –$\frac{1}{8}$ in.) | 1·6–3 mm ($\frac{1}{16}-\frac{1}{8}$ in.) | 0·8–3 mm ($\frac{1}{32}-\frac{1}{8}$ in.) | | 100–127 | 2·4 mm ($\frac{3}{32}$ in.) |
| | | | | 90–100 | 3 mm ($\frac{1}{8}$ in.) |
| 3–5 mm ($\frac{1}{8}-\frac{3}{16}$ in.) | 3–3·8 mm ($\frac{1}{8}-\frac{5}{32}$ in.) | 80°V  1·6–3 mm ($\frac{1}{16}-\frac{1}{8}$ in.) | | 75–90 | 4 mm ($\frac{5}{32}$ in.) |
| | | | | 60–75 | 4·8 mm ($\frac{3}{16}$ in.) |
| 5–8·2 mm ($\frac{3}{16}-\frac{5}{16}$ in.) | 3–3·8 mm ($\frac{1}{8}-\frac{5}{32}$ in.) | 3–3·8 mm ($\frac{1}{8}-\frac{5}{32}$ in.) | | 50–60 | 6·4 mm ($\frac{1}{4}$ in.) |
| | | | Rightward welding | 35–40 | 8 mm ($\frac{5}{16}$ in.) |
| 8·2–15 mm ($\frac{5}{16}-\frac{5}{8}$ in.) | 3·8–6·5 mm ($\frac{5}{32}-\frac{1}{4}$ in.) | 60°V  3–3·8 mm ($\frac{1}{8}-\frac{5}{32}$ in.) | | 30–35 | 9·5 mm ($\frac{3}{8}$ in.) |
| | | | | 22–25 | 12·5 mm ($\frac{1}{2}$ in.) |
| 15 mm ($\frac{5}{8}$ in.) and over | 6·5 mm ($\frac{1}{4}$ in.) | Top V 60° / Bottom V 80°  3–3·8 mm ($\frac{1}{8}-\frac{5}{32}$ in.) | | 19–22 | 15 mm ($\frac{5}{8}$ in.) |
| | | | | 15–16 | 19 mm ($\frac{3}{4}$ in.) |
| | | | | 10–12 | 25 mm (1 in.) |

Figure 3.14 Oxy-acetylene process variables (data courtesy of B.O.C. Ltd)

Table 3.2 Shades of glass for eye protection

| Grade | | | |
|---|---|---|---|
| Gas welding without flux | Gas welding with flux | Shade number | Typical applications |
| GW1 | GW1,F | 3 | Gas welding of aluminium and magnesium—aluminium alloys; lead burning or oxy-acetylene burning |
| GW2 | GW2,F | 4 | Oxygen machine and hand cutting; oxygen gouging; flame de-scaling; silver soldering; fusion welding of zinc-base die-castings; bronze welding of light-gauge copper pipe and light-gauge steel sheet |
| GW3 | GW3,F | 5 | Fusion welding of copper and copper alloys; fusion welding of nickel and nickel alloys; fusion welding of steel plate; all bronze welds in heavy-gauge steel and cast iron, except pre-heated work; rebuilding work of relatively small steel parts and areas for fusion welding; all hard-surfacing operations including rail resurfacing |
| GW4 | GW4,F | 6 | Fusion welding of heavy steel; fusion welding of heavy cast-iron; fusion welding and bronze welding of pre-heated cast iron and steel castings; rebuilding large steel areas, e.g. large cams, etc. |

*Figure 3.15    Clamping with shaded pole magnets (courtesy of James Neill and Co. Ltd)*

A useful formula for relating filler-rod diameter to work-piece thickness (square butt preparation) is

$$D = \tfrac{1}{2}T$$

where $D$ is the rod diameter and $T$ the plate thickness.

Conditions for welding various material thickness are shown in table 3.1 (this table applies only to equal-pressure or high-pressure equipment).

A desirable feature of an oxy-acetylene torch is that the control-valve knobs are situated conveniently for the thumb and revolve easily under light pressure. This enables the operator to adjust the flame while working — the position of the operator's thumb can clearly be seen in figure 3.15 (note also the use of magnets for locating and holding the tube sections).

It is important to use the correct shade of welding glass in the goggles used for welding and brazing, fluxes containing sodium and potassium salts and some metallic oxides like those of aluminium and magnesium, emit a strong light when heated and require the use of glasses according to table 3.2.

## CAST IRON

There are three common varieties of cast iron: *grey, white* and *malleable*. Another type finding increasing use is S.G.I., or spheroidal graphite cast iron.

Grey cast-iron is manufactured by slowly cooling liquid iron containing about 3 per cent carbon, the grey colour of the fractured metal being due to free carbon in the form of graphite. White cast-iron reveals a lighter colour when broken and is produced by the rapid cooling of the metal from liquid; it also has a carbon content of approximately 3 per cent and consists, in the solid, of aggregates of pearlite in cementite. Malleable cast-iron is made by heating, to about 850 °C for several days, iron that is low in carbon and high in silicon. Pulverised iron-ore, mill scale and sand are added and the annealling treatment promotes breakdown of the cementite into discrete particles of evenly distributed graphite (temper carbon). Some of the carbon is concentrated near the surface of the material and becomes oxidised, forming a layer of iron of low carbon content (steel). The element silicon occurs in cast iron in amounts between 0.5 and 3 per cent, the usual figure being between 1.5 to 2 per cent. Silicon promotes the dissolution of cementite and the change to graphite, while the addition of manganese increases the stability of the cementite, making the casting stronger and harder.

### Preparations for Welding Cast Iron

Broken sections should have a single-vee preparation, 90° inclusive angle, in material up to 12 mm thick. Above this thickness a double-vee preparation is necessary. The parts should be carefully aligned using fireclay, compound asbestos or carbon paste. If possible, the work should be pre-heated in a furnace, but oxy-acetylene or propane torches can be used, in which case care must be taken to heat the parts uniformly. When the temperature of the work has reached 500 to 550 °C (dull red) welding is begun.

The filler rod is usually in cast form (about 6 mm square section) and is low in carbon and high in silicon. The use of a flux is necessary to dissolve the oxides formed during the heating and melting of the parent material. Fluxes for this purpose contain balanced proportions of alkaline carbonates, bicarbonates and borates.

### Welding Procedure

The leftward technique is used, keeping the torch at a high angle, see figure 3.16. The flame force is used to control the extremely fluid metal; the neutral flame must be used and it is important to ensure that it does not become oxidising. The cone should not be allowed to contact the weldpool, otherwise hard spots may result. A cone-to-work distance of about 8 mm should be maintained but in cases where the cast iron exhibits

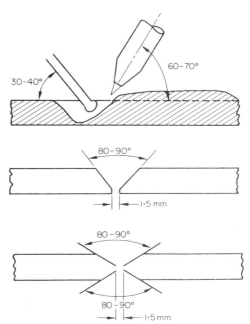

*Figure 3.16    Use of filler rod*

a lack of fluidity, a very slight excess of acetylene may improve matters.

Slow cooling of cast iron is essential; for this purpose a muffle may be used or the weldment can be immersed in dry sand, wrapped in asbestos or covered with pulverised fire-bricks.

### BRAZE WELDING

#### Braze Welding of Cast Iron

This is mainly a repair and reclamation procedure; it has the advantage that pre-heating time can be reduced because of the lower fusion-temperature required, and certain repairs can be carried out *in situ*. Uniform pre-heating to about 400 °C is necessary for heavy sections, followed by a slow cooling to normal temperature.

A neutral flame is first obtained — and then further adjusted to give a slight excess of oxygen — see figure 3.17. The leftward method is used for welding; the work can be slightly inclined to give better weldpool control.

#### Braze Welding of Copper

Tough pitch or non-deoxidised copper can be joined to itself or to certain other metals by braze welding. Some joint designs

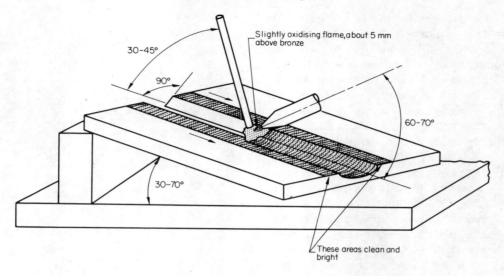

*Figure 3.17    Bronze welding cast iron*

for copper-pipe welding are shown in figure 3.18. Joints made in copper tubes are usually stronger than the parent material. Pre-heating is only necessary on large-diameter pipes; the flame should be very slightly oxidising and the leftward-welding technique used.

### Tin and Lead Bronzes

Fusion is difficult with these alloys unless a strongly oxidising flame is employed with a matching filler. Owing to the large number of these alloys, great care should be taken in selecting the correct filler-material (manufacturers will usually supply this information).

### Aluminium Brazing

This process is still widely used in the manufacture of small vessels, domestic equipment, radio chassis, etc., but it has largely been superseded by T.I.G. A slightly carburising flame is necessary, with just the suspicion of an acetylene haze at the end of the inner cone. A continuous check should be made to maintain this condition when brazing since the tendency of the flame is to become oxidising.

Joint fit-up should be good without being an interference fit, that is, slight clearance is required so that the brazing alloy can be drawn between the joint faces by capillary attraction. Joints suitable for aluminium flame brazing are shown in figure 3.19. Interface depths should be kept to a minimum, as

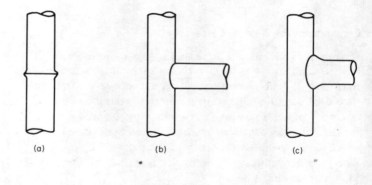

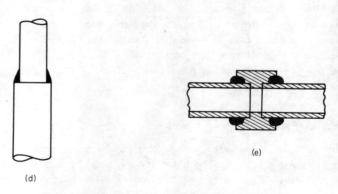

*Figure 3.18    Copper-pipe joints for bronze welding; (a) bell butt, (b) branch tee saddle, (c) bell tee, (d) diminishing, (e) standard socket*

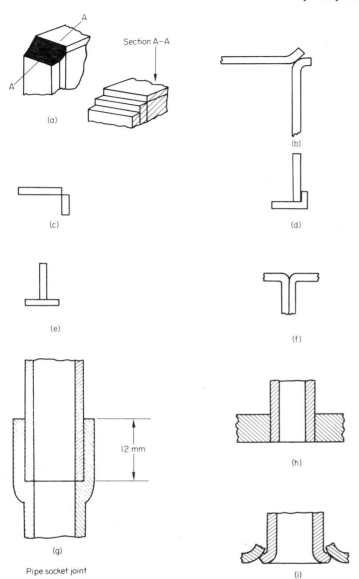

Figure 3.19    Joints for aluminium brazing

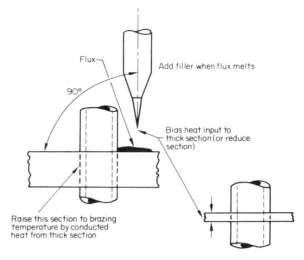

Figure 3.20    Brazing unequal sections

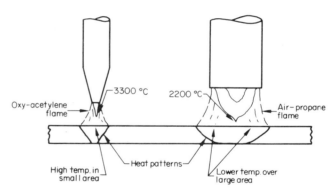

Figure 3.21    Comparison between air—propane and oxy-acetylene flames

shown in figure 3.19c. Careful torch-manipulation is required for assemblies of different section, figure 3.19h. In light-gauge materials joints (b), (c), (d), (e), (f) and (i) are suitable, but if a multi-sheet assembly is necessary then stepping of the edges, as in figure 3.19a, section A—A, can be used. If the difference between the sections is extreme, as in figure 3.20, great care is needed in pre-heating; preference must be given to the thick member, heat being allowed to flow into the thin member only by conduction, since direct heating is likely to result in collapse.

An alternative to oxy-acetylene is the air—propane flame. This has several advantages, particularly for unskilled operators (see figure 3.21). Because the flame has a large cross-sectional area it protects the metal adjacent to the joint from atmospheric attack and promotes the spread of flux. The flame temperature is about 1000 °C lower than oxy-acetylene, and so reduces the risk of parent-metal collapse; in addition, the flame can be held steadily over the joint, unlike the oxy-acetylene flame which, because of its high temperature developed in a small area, requires constant manipulation.

Commercially pure alumunium is suitable for brazing, but aluminium with a magnesium content in excess of 2 per cent is unsuitable. Very thorough cleaning of the joint interfaces and local areas is necessary. Steel wool, a wire brush, or a clean file may be used and care must be taken not to handle the prepared surfaces afterwards. *Trichlorethylene or carbon tetrachloride should not be used* for degreasing because of their dangerous toxic properties. Flux residues must be removed by scrubbing in hot soapy water followed by a rinse in clean water.

### Low-temperature Brazing

This is also known as hard soldering or silver soldering and the alloys used have a melting range from 600 to 850 °C. Most of these alloys are based on silver, zinc and copper, whereas the high-temperature brazing or 'bronze' welding alloys are based on copper and zinc and may also contain manganese, silicon, nickel or tin. Whereas the purpose of high-temperature brazing is to achieve strength mainly by deposit build-up (except on sheet metal where capillary conditions exist more often) the silver/copper alloys are characterised by their high capillary-flow.

Silver soldering is widely used for joining steel and copper and its alloys, as well as dissimilar metal combinations. Cadmium is sometimes added to promote fluidity, but use of materials containing cadmium should be avoided wherever possible owing to the toxic character of this metal. A good substitute is an alloy containing a higher proportion of silver, but because of the extra cost, sparing use should be made of it.

Joint design for silver soldering should be precise (between 0.2 and 1 mm for steel, but this varies with material type). Unless this is achieved the advantage of capillary flow will be lost and the molten alloy may escape from the joint. Adequate cleaning of the joint surfaces is very important to obtain good wetting without overheating the alloy or flux. The flux usually contains a proportion of fluoride salts and the fumes must be avoided by correct heating and ventilating of the workpiece.

After silver soldering, work should be washed in hot water or boiled to remove flux residues. Automatic silver-soldering is widely used for thin tubular components, the alloy, in the form of a ring, being placed around the joint and heat provided by electrical induction. As with aluminium brazing, care must be taken to heat workpiece sections uniformly, taking special precautions to do so if there is a wide difference of gauge.

## HARDFACING

The oxy-acetylene flame can be used for depositing an overlay of the requisite hardness in four distinct ways

1. By using an appropriate filler material in rod form, with or without flux.

2. By injecting a finely divided alloy metal powder into the gas stream of the torch.

3. By controlled backfiring of the flame into which a finely divided metal powder has been introduced (this process is known as flameplating and the service is provided in the United Kingdom by the Union Carbide Corporation).

4. Metal spraying, in which alloy or elemental metals are fed into an oxy-acetylene flame and disentegrated by a carrier gas, usually compressed air.

For wide application the technique of using a filler rod manually is employed — see figure 3.22. Special alloy filler rods are necessary, among the most commonly used being the stellite alloys, based on cobalt and nickel. A carburising flame is required for depositing hard alloy material, thus producing, initially, a carbon-rich sweating layer on the workpiece. Controlled pre-heating and cooling are necessary when hardsurfacing carbon and alloy steels and the thickness of the deposit should be kept to a minimum. Cast iron cannot be hardsurfaced easily because it does not exhibit the sweating characteristic of steel.

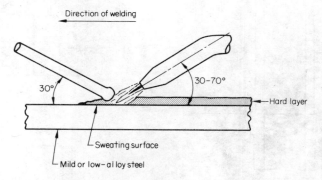

*Figure 3.22    Hardsurfacing technique*

Tungsten carbide in a matrix of nickel and cobalt gives a high-hardness layer, but is not machinable by normal methods and alloys that develop their full hardness after heat treatment are often used. The method of hard-coating by using metal powder inducted from a small reservoir attached to the torch is used where comparatively thin deposits are required, that is, up to about 0.8 mm thick.

Using the flameplating system, even greater accuracy of deposit thickness can be attained, and there is the further advantage that heat accumulation in the workpiece is substantially reduced — an important consideration when dealing with

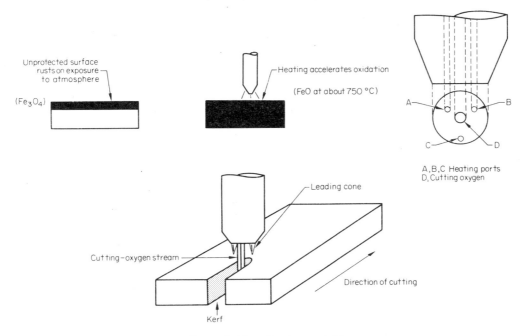

*Figure 3.23    Cutting variables*

high-carbon heat-treatable steels. Flame spraying is a very versatile technique with almost limitless uses. In addition to providing hard coatings to metals, the process can be used for applying metals or ceramics to non-metals like wood, brick and plastics. It is extensively used in the chemical-plant industry for coating vessels with anti-corrosion materials, similarly for protecting ships, buildings, bridges, etc., from their environments.

## FLAMECUTTING

The natural rusting that occurs when iron is exposed to the atmosphere can be compared to the accelerated action provided by oxygen on the molten metal. The flamecutting torch consists essentially of a means of pre-heating the metal to the required ignition-temperature and a way of controlling the cutting oxygen. Originally a cutting torch had a completely circular heating-flame, but this was abandoned in favour of separate heating-jets situated at equal distances around the tip, as shown in figure 3.23. The number of heating ports may vary according to the heat input required. The bore of the centrally situated oxygen-cutting stream also varies with the thickness of material to be cut.

For quick starting, oxy-acetylene is preferable, but on heavy-gauge plate a rounded kerf-edge is likely, a feature not

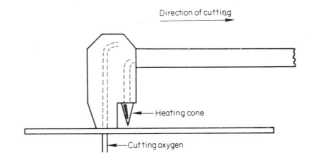

*Figure 3.24    Step-cutting head for light-gauge materials*

often desirable. For this reason, alternative fuel-gases giving lower flame-temperatures may be used. Among these are coal gas, natural gas and propane. Thin material is more efficiently cut using one of these gases, but a special oxy-acetylene torch (shown in figure 3.24), called a step cutter, is also available and enables high-speed narrow cuts to be made in steel down to 0.8 mm thick.

The high consumption of oxygen in flamecutting necessitates the use of special cylinder pressure-regulators and an adequate supply of oxygen. This can be provided from a cylinder manifold or from a bulk-liquid supply; the cylinder-pressure regulator for oxygen flamecutting should have a capacity of up to 1.05 MN/m² (10.5 bar), and for acetylene up to 0.1 MN/m² (1 bar).

## Starting a Cut

The edge of a plate is the usual site for starting a cut (see figure 3.25), but this can often be wasteful, particularly if both parts of the cut plate are required for use, as in producing press tools or metal-working tools. For this reason a cut must sometimes be started on the cut line as shown in figure 3.26. Machine cutting is superior in every way to hand cutting — limits of ± 0.8 mm can easily be achieved.

In figure 3.27 the principle of the flamecutting pantograph is seen together with the tractor-mounted cutter for straight-line or circle cutting. For hand operation a trammel bar or

guide block is useful, but practice in assessing the required travel-speed for a specific plate-thickness is essential. This particularly applies to hand bevel preparations of thick material, and for accuracy these should be executed by machine.

The affinity of oxygen for ferrous metals, especially when these are heated to, or above, their ignition temperature, is the basis of the oxy-acetylene flamecutting process. This can be illustrated by the classical experiment where a heated watch-spring is burnt in a jar of oxygen. Perhaps a more practical way to demonstrate the principle is as follows. If a thin steel sheet

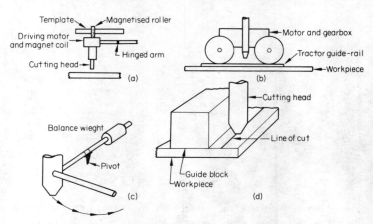

Figure 3.27  Cutting methods; (a) pantograph cutting, (b) tractor-drive cutting, (c) trammel bar for circle cutting, (d) cutting using guide bar

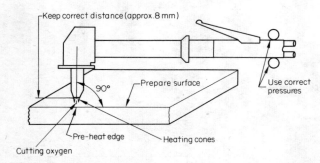

Figure 3.25    Starting a flame cut

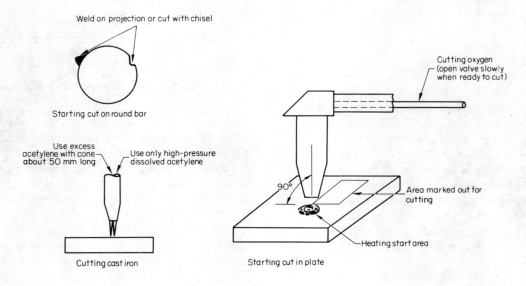

Figure 3.26    Cutting cast iron

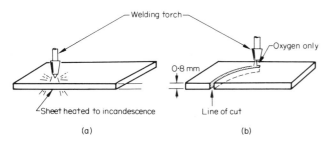

*Figure 3.28    Iron—oxygen reaction*

of, say, 0.8 mm thickness is strongly heated with an oxy-acetylene torch (see figure 3.28a), until the area beneath the flame starts to melt, the acetylene supply can be quickly shut off and the metal cut solely by the stream of oxygen, as in figure 3.28b. The chemical reaction responsible for the flame-cutting of ferrous metals is started when the area beneath the heating flame reaches about 870 °C. At this point, a stream of pure oxygen issuing from the centre port of the nozzle causes rapid oxidation. The torch is then moved, by hand or machine, to produce a cut or kerf. Because the reaction takes place only within the area covered by the oxygen jet, extremely accurate control of the kerf width is possible.

Besides acetylene, other fuel gases can be used such as propane, coal gas, natural gas, and mixtures of hydrocarbons derived from petroleum distillation. It is necessary to use only the cutting head or nozzle for each fuel gas as recommended by the manufacturer.

Machine flamecutting enables very precise profiling to be made, for example, sprockets for armoured vehicles have been cut to a tolerance of ± 0.14 mm.

## FLAMEHARDENING

The principal considerations in the flamehardening of metals are the cooling rate specified for the material, and the alloy content. Since carbon is the main constituent of hardenable steels it is of major importance in the flamehardening process. In addition to carbon, other elements like manganese, chromium, vanadium, silicon, nickel and molybdenum are often present. Flamehardening is an advantage because it can be accurately controlled to give precise depth of hardening and is therefore used where hard surfaces capable of heavy wear and abrasion are required.

There are four ways in which flamehardening can be used

1. The work is indexed beneath the heating flame which is then shut off and a cooling spray applied to the work-piece.

2. One or more heating flames are situated so that the work passes continuously beneath them.

3. The heating flames are traversed automatically over the stationary workpiece.

4. The heating flame travels horizontally or longitud-inally across the rotating workpiece.

Method 1 is called consecutive hardening and methods 2, 3 and 4 are called progressive hardening.

Oxy-acetylene heating, using a neutral-flame condition, is most often used because of its excellent control and heat-transfer characteristics. Fine-grain steels are the most suitable for flamehardening because they tend to resist grain growth during heating.

## FLAMESPRAYING

Metals, like copper, zinc, lead, stainless steel, brass and aluminium in the form of wires can be used to coat metal or non-metal surfaces, by the process known as flame-spraying. A specially designed torch comprising a burner for oxygen and acetylene, wire-feed conduit and air-feed tube is required. The emerging wire is melted by the oxy-acetylene flame and projected by the air blast in the form of finely divided particles. Before flamespraying, the work must be degreased and shot-blasted to provide a keying surface. Metal deposited is only mechanically bonded, for fusion an arc process must be used.

*Note.* Plasma-arc spraying is used for high-melting-point materials like alumina (2030 °C), cerium oxide (2670 °C), hafnium oxide (3520 °C), magnesium zirconate (2130 °C), tungsten (3370 °C), tungsten oxide (1925 °C), tungsten carbide (2850 °C), zirconium oxide (2700 °C) and mixtures like tungsten carbide and cobalt.

For small-scale oxy-acetylene flamespraying, finely divided metal-powders are used by inducting them into the gas stream of the torch from a small reservoir. This is an extremely useful technique and finds many uses in jig and tool manufacture, press-tool repair, etc.

## FLAMEGOUGING

A specially designed oxy-acetylene torch is used for this process, which is a relatively cheap way of removing defective surface-areas of metal, welding-bead defects and the preparation of plates for welding. The flamegouging torch is designed to deliver a large volume of oxygen at low nozzle-velocity coordinated with appropriate heating-cone distribution.

The main method of operation is spot gouging. The defective area is marked out (for example, by centre punch dots) and the pre-heating cones are adjusted to give slightly oxidising flames. A starting point is chosen a little to one side of the marked area. When the surface is suitably heated, the cutting oxygen is turned on and the angle of the torch is increased towards the vertical so that the cutting-oxygen jet is increasingly downwards. The heating-flame cones should be maintained throughout at a distance not exceeding 2 mm from the work surface.

## FURTHER STUDY

A. Dickason, *Technology of Sheet Metalwork*, Pitman, London, 1967

F. Koenigsberger and J. R. R. Adair, *Welding Technology*, Macmillan, London, 1966

D. Romans and Eric Norman Simons, *Welding Processes and Technology*, Pitman, London, 1968

B. E. Rossi, *Welding Engineering*, McGraw-Hill, Maidenhead, 1954

B.O.C. Ltd, *Acetylene, its Properties, Production and Uses*, Waltham Cross, 1969

B.O.C. Ltd, *Handbook of Operating Instructions*, Waltham Cross, 1970

B.O.C. Ltd, *Production of Oxygen, Nitrogen and the Inert Gases*, Waltham Cross

B.O.C. Ltd, *Safety in the use of Compressed Gas Cylinders*, Waltham Cross

Department of Employment and Productivity, *Repair of Drums and Small Tanks*

Welding Institute, *Health and Safety in Welding*, Cambridge

H.M.S.O. Booklet 3 (New series), *Bulk Liquid Storage of Liquefied Gas Cylinders*

H.M.S.O. Booklet 34 (New series), *Guide to the Use of Flame Arresters and Explosion Reliefs*

*H.M.S.O. Forms 386 and 1926*

*H.M.S.O. Safety booklet 18*

**Books and instruction manuals**

C. G. Bainbridge and F. Clarke, *Oxy-acetylene Welding Repair Manual*, B.O.C. Ltd, Waltham Cross, 1968

Ethel Browning, *Toxicity of Industrial Metals*, Butterworth, London, 1969

G. Love, *The Theory and Practice of Metalwork*, Longman, London, 1967

B.O.C. Ltd, *Handbook for Oxy-acetylene Welding*, Waltham Cross, 1967

B.O.C. Ltd, *Rightward Welding Booklet 1*, Waltham Cross, 1967

B.O.C. Ltd, *Vertical Welding Booklet 6*, Waltham Cross, 1967

British Acetylene Association, *Portable Apparatus for Welding*, London

Lincoln Electric Co., *Metals and How to Weld Them*, Machinery Publishing, Brighton

Welding Institute, *Brazing, Soldering and Braze Welding*, Cambridge

H.M.S.O. Booklet 35 (New series), *Basic Rules for Health and Safety at work*

H.M.S.O. Form 814, *Memorandum on Explosion and Gassing Risks in the Examination and Repair of Tanks and Stills*

BS 349:1973 Identification of contents of industrial gas cylinders

BS 638:1966 Arc welding plant, equipment and accessories

BS 679:1959 Filters for use during welding and similar industrial operations

BS 1389:1960 Dimensions for hose connexions for welding and cutting equipment

BS 1547:1959 Flameproof industrial clothing (materials and design)

BS 1821:1957 Class I oxy-acetylene welding for steel pipelines and pipe assemblies for carrying fluids

BS 2640:1955 Class II oxy-acetylene welding of steel pipelines and pipe assemblies for carrying fluids

BS 2653:1955 Protective clothing for welders

*The Gas Cylinders (Conveyance) Regulation*, S., R. and O., 1938–654

*The Petroleum (Carbide of Calcium) Order*, S., R. and O., 1947–1442

# 4. GAS-SHIELDED METAL-ARC WELDING

When a bare-wire consumable electrode fed from a reel is used with a chemically inert gas such as argon or helium, the process is called M.I.G. (metal—inert-gas). If chemically active gases like carbon dioxide, hydrogen, nitrogen or oxygen are used, the process is called M.A.G. (metal—active-gas).

Argon and helium are elements that do not enter into chemical combination with other elements (hence 'inert'), and are therefore ideal in many cases for arc shielding, especially when welding reactive metals like aluminium, copper, nickel, chromium and its alloys, etc.

Carbon dioxide, $CO_2$, is a chemical compound of the elements carbon and oxygen; the gas is non-reactive with metals except at arc temperatures, when it decomposes to produce carbon monoxide and free active oxygen. The elements hydrogen and nitrogen do not attack ordinary metals in their solid state but they may do so when the metals are strongly heated. There are exceptions to this — for example, copper, which is not attacked by nitrogen and can be T.I.G. welded with that gas. Oxygen is extremely reactive with many metals at normal temperatures. When combined with hydrogen to form water it can cause serious porosity problems if drawn into the arc atmosphere.

Oxygen, hydrogen, nitrogen and carbon form many organic compounds of which greases and lubricants are common examples. These substances, known generally as hydrocarbons, can also produce weld contamination and porosity if introduced via the arc. The importance can now be seen of thoroughly cleaning the workpiece before welding.

G.S.M.A. (gas-shielded metal-arc) welding using carbon dioxide as a shielding gas is particularly sensitive to water, liquid or vapour, and more tolerant of hydrocarbon compounds (paraffin, lubricating oils, etc.) than, say, T.I.G. Work coated lightly with thin rust-preventing oil is satisfactorily welded by G.S.M.A. using $CO_2$ shielding.

Some of the other names used to describe G.S.M.A. welding are 'short-arc' 'semi-automatic' and '$CO_2$' welding.

The G.S.M.A. process (using a small-diameter spool-wound electrode fed into an arc protected by a gas shield) first came into prominence industrially around 1950. Before then aluminium and its alloys could only be welded by metal-arc and the oxy-acetylene flame. Both of these techniques, however, required the use of highly active fluxes that left corrosive residues. Fortunately T.I.G. (tungsten—inert-gas) welding appeared at about the same time and proved superior in many ways, in addition to being a flux-free process.

In the early days of G.S.M.A. welding, argon with about 1—2 per cent oxygen was used and because of the high cost of argon a search was made for a less costly gas. Carbon dioxide, being cheap and easy to manufacture, indicated some merit as an arc-shielding gas for bare-wire welding. Unfortunately, because the gas progressively decomposed with increase in temperature, there was the problem of how to combat the carbon monoxide and oxygen thus produced. Both of these gases, if present in the arc, react with molten metal to produce porosity and contamination. The $CO_2$ shield, however, despite its chemical breakdown under arc heat, did prevent attack by external (to arc) atmosphere and gases like water vapour, oxygen and nitrogen. Carbon monoxide and oxygen generated from carbon dioxide were no longer a problem when de-oxidizing elements (that is, elements that have a strong affinity for oxygen) like manganese, silicon, titanium, aluminium or zirconium were added to the electrode material. Not all of these elements are commonly used — silicon, manganese and aluminium are the main additives.

Figure 4.1 shows the essential principles of the G.S.M.A. process. (It is worth noting here that the flux in the metal-arc welding generates carbon dioxide as an arc shield but deoxidizing elements are also present.)

In metal-arc welding the capacity of any given diameter of electrode to pass large increases in current is severely limited. For example, a 2 mm diameter electrode can be used in the range from 50 to 80 A but if the current was increased to, say, 170 A it would be necessary to use an electrode of 4 mm diameter. In the G.S.M.A. process this change of electrode diameter with increase in current is not necessary over a wide range; for example, 0.8 mm diameter wire would be used quite successfully in the range 50 to 200 A.

It can be said that the speed at which the wire is supplied to

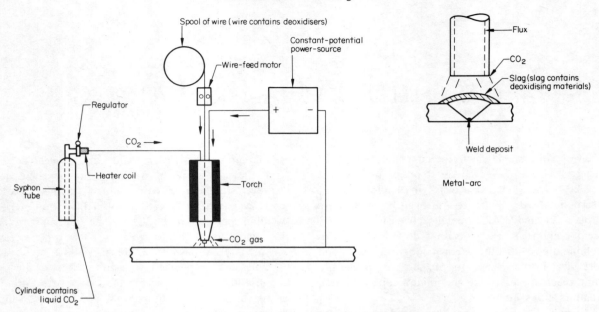

*Figure 4.1    Gas-shielded metal-arc welding*

the arc and the current required for melting have a practically linear relationship which is known as the burn-off rate. This rate is related to electrode diameter and composition, that is, it would vary for steel, aluminium or stainless steels, for example. To supply more current it is necessary to increase the wire speed. This offers a great advantage because it means that when the correct wire-speed according to the burn-off rate has been selected, the current level is automatically correct. To obtain these output characteristics a power source with a slightly drooping constant-potential output is used. If we compare the drooping characteristic of a metal-arc power-source we can understand that this provides a high open-circuit voltage and constant-current output. This is suitable for metal-arc welding because it provides a condition of relative insensitivity to electrode movement (that is, variations in arc length). With a constant-potential power-source, however, a small change in arc length affects the current supplied so that the arc length is rapidly readjusted, hence the term 'semi-automatic' welding.

## METAL TRANSFER IN G.S.M.A. WELDING

By metal transfer we mean the manner in which metal passes from the electrode to the weldpool or workpiece immediately beneath the arc. Using carbon dioxide as a shielding gas, there are two types of transfer, (i) free-flight or spray, and (ii) dip or

short-circuit. Type (ii) is the most common mode, true spray-transfer only occurring in $CO_2$ at high voltages. Type (i) can be used with low-voltage values if an argon-rich gas (80 per cent A + 20 per cent $CO_2$) is used.

For the welding of steel up to about 5 mm thick the dip-transfer system is satisfactory. Heavier-gauge material requires greater heat-input – a characteristic of the spray mode. At low welding-currents, metal is transferred as large non-axial droplets (that is, not falling along the perpendicular axis of the arc). Dip-transfer conditions are attainable in the 16 to 32 V output range. These, in increments of 2 or 3 V, are usually indicated on the operating panel of the power source. The free-flight-transfer condition overlaps this, starting at about 26 V and extending for practical purposes to about 42 V.

When welding commences, using the dip-transfer mode, the electrode short-circuits to the workpiece and the current rises, burning off the end of the wire. This is followed by the re-establishment of the arc. Immediately a droplet forms on the electrode and is carried forward to complete the next short-circuit. To control the short-circuit current rise-time, an inductor is used. This is usually situated in the front of the welding set, in some cases pre-set tappings are used, provided with plug-in sockets.

In the welding of light-gauge steel, say, 2 to 5 mm thick, the short-circuit time is ideally the same as the arc-duration

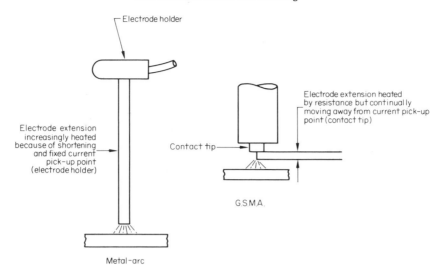

*Figure 4.2    Electrode heating-effect*

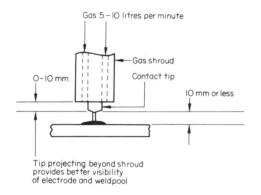

*Figure 4.3    M.I.G. torch details*

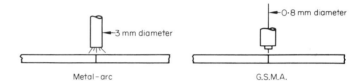

*Figure 4.4    Electrode comparison*

time and so very little inductance (or choke) is required. The rate of short-circuits in this range lies between 100—130 per second but with argon-rich gases, up to 150 short-circuits per second are attainable. Generally increasing the inductance also increases heat input to the workpiece but excessive inductance will cause spatter and arc instability. The 'tuning' of the power-source output comes with experience and when conditions are correct the arc has a characteristic ripping or crackling sound. It is important to remember that the inductance value also varies according to wire diameter and composition.

Figure 4.2 compares the resistance-heating effect on the electrodes when using metal-arc and G.S.M.A. The electrode

extension in the latter is called the wire stick-out and is an important control medium in manual welding since it can be used to modify heat input to the workpiece. The contact tip is the nozzle or tube at the extremity of the welding gun; this is where the moving electrode picks up the current supplied from the power source — see figure 4.3. Improved work-area visibility results if the contact tip projects, as shown, beyond the gas shroud.

Much greater tracking skill (that is, following the joint line accurately) is necessary with the small-diameter electrodes used in G.S.M.A. welding than with metal-arc. Figure 4.4 shows the difference in the relationship of the electrode diameters to the joint line.

## SAFETY

Before using G.S.M.A. equipment there are certain rules of safety that must be observed and these should be studied well.

Remember that all cylinders of compressed gas are potentially dangerous if wrongly handled or situated (near heat sources or in places where they may accidently form part of the welding-current return-path, etc.). Carbon dioxide is supplied in liquid form in high-tensile-steel cylinders. These have a special threaded gland outlet and no attempt should be made to fit other than the correct pressure-regulator.

Carbon dioxide remains liquid under pressure in the cylinder. When the outlet valve is opened, liquid is drawn from the bottom of the cylinder through the syphon tube as was shown in figure 4.1. Carbon dioxide cannot exist as a liquid at atmospheric pressure — under these conditions it changes first to a solid ('dry ice') and then to a gas. When the liquid begins to flow out of the cylinder the transition from liquid to gas is maintained by using a thermostat-controlled heater-coil. If this were not used, the regulator would quickly become blocked with solid $CO_2$.

Always open the cylinder valve slowly, since a sudden surge of pressure may damage the regulator. Make sure all pipe connections are tight — a small leak will cause local freezing. *Use only soapy water for leak testing.* Do not open a cylinder valve unless a regulator is fitted. Keep the cylinder upright when in use by securing it in a properly designed truck or stand. Always close the cylinder main valve before disconnecting the regulator. Use only the correct outlet-fittings on cylinders in accordance with BS 341: Valve fittings for compressed gas cylinders Part 1:1962, Valves with taper stems, outlet connection No. 8. Never use Stillsons or pipe wrenches on connections, use only the correct-size spanners.

The electrode wire emerging from the welding gun, when the start switch is operated, can be dangerous — instances have been known where the wire has penetrated the welder's hand — *use gloves.*

It is most important that a good welding return-connection is made between the workpiece and return cable, the interfaces of the connecting spade-terminal and work must be clean and the joint tight.

In confined spaces such as tanks, vessels, etc., be sure that there is good ventilation and check equipment (pipes, joints, union nuts, etc.) for leaks, remembering that carbon dioxide can cause death by suffocation. (The gas is slightly heavier than air and therefore tends to collect at low level.) Check for equipment leaks with soapy water only.

Cylinders are provided with bursting discs; these are safety devices for releasing excess pressure which, for example, could occur if the cylinder was heated. Never attempt to renew a burst disc, but mark the cylinder for immediate return to the supplier.

In some instances, carbon dioxide is supplied by pipeline at considerably lower pressure than that in the cylinder, which is about $6 MN/m^2$. Line pressure, which may be up to $0.7 MN/m^2$, can be equally dangerous if mishandled. Always use the correct fittings when connecting supply pipes between the line and the welding set.

## PREPARATION FOR WELDING

Assuming that the equipment is connected to the power supply and the valves controlling the flow of carbon dioxide have been set, there are three controls to be adjusted; these are voltage, inductance ('reactance' or 'choke'), and wire speed.

The control of welding current is achieved by varying the wire speed, and because the power source has a flat characteristic, the open-circuit voltage is very nearly equal to the voltage across the arc (arc voltage).

An average gas-flow for light-gauge metal fabrication would be 10 litres/min, but this may have to be increased in draughty conditions or modified according to the gas-shroud diameter and the wire extension (stick-out), or the shape of the workpiece. It is not good practice to use excessive gas-flow.

Satisfactory arc-shielding is indicated by the slag islands (glass-like deposits) that can be seen on the deposited metal. These are produced by the chemical action of the deoxidant materials contained in the electrode reacting with oxygen and carbon monoxide produced by $CO_2$ breakdown in the arc. If there is insufficient gas-protection, atmospheric gases (mostly oxygen, nitrogen and water vapour) attack the molten weld-metal because there is not a sufficient quantity of deoxidisers in the wire to cope with such conditions.

Since the inductor controls the response of the rectifier to the electrode short-circuits, adjustment will have to be made to produce a condition of metal transfer that will give a satisfactory welding-condition.

If 0.8 mm wire is used, a typical welding condition would be 20 V (open-circuit), welding or arc voltage about 18 V, wire speed about 5 m/min (about 200 in./min), current 115 A. This setting would be suitable for welding steel sheet in the range of 0.6 to 1.6 mm thick.

Another example would be for welding 3 mm thick sheet (with the same electrode-diameter): open-circuit voltage 22 V, arc voltage about 20 V, wire speed 5.5 m/min, welding current 125 A.

### Effect of Wire Extension on Deposition

A lengthening of the electrode extension from the contact tip to the arc will increase the electrical resistance and result in further heating of the wire at this point. Similarly if the electrode diameter is decreased (without any other alteration to the output parameters) an increase in electrical resistance will result. The outcome of these changes is seen in the increased droplet-size on the electrode tip and the lower frequency of short-circuits. Withdrawing the torch from the work will promote self-adjustment of the arc but the lengthened wire-extension will be correspondingly resistance-heated.

The wire extension is a useful control medium for the manual welder because it offers the opportunity of adjusting heat input to suit local joint-geometry. For example, by drawing the torch a little further from the joint when a gap condition exists, the droplet diameter on the electrode will increase and the short-circuit frequency will decrease. This will enable the gap to be bridged more easily.

## G.S.M.A. WELDING PRACTICE

The leftward-welding technique (figure 4.5) is most often used. This produces a more rounded and somewhat flatter weld-bead than the rightward method. This is because using the leftward progression, the arc tends to be in advance of the weldpool and thus pre-heats the forward area; this means that there is a slightly longer cooling-period from liquid to solid weld metal.

With the rightward system, the arc is biased towards the weld deposit and so there is less heating of the joint area

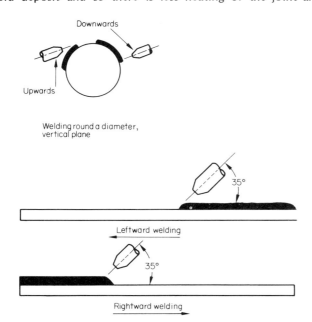

*Figure 4.5     Weld directions*

immediately in front of the weldpool. Welding around a radius (figure 4.5) is easier if the progression is downwards rather than upwards, but there are exceptions, such as welds in thick material, when the upwards technique is often necessary.

The main range of application of the process in the welding of steel and alloys lies in the thickness between 0.6 and 50 mm, but it is in the fabrication of sheet-metal goods that G.S.M.A. is particularly useful. Formerly oxy-acetylene and arc welding with flux-coated electrodes were used. The distortion of the work by oxy-acetylene welding is to a large extent inevitable, but with G.S.M.A. this is substantially reduced because of the high-intensity local heat of the arc, the fast metal-deposition rate and speed of travel.

Figure 4.6a shows an edge-to-edge corner joint in 0.8 mm steel. A steel alignment-block with a bevelled edge is used for holding the parts by clamping. This arrangement would be suitable for oxy-acetylene welding but if the copper block in figure 4.6b were substituted it would be extremely difficult to obtain melting of the sheet edges or to establish a weldpool. The reason for this is the high thermal-conductivity of copper. If G.S.M.A. were used, however, the substantially higher temperature of the arc and the high deposition-rate would overcome the problem. Although there is a gap, which could be up to three times the metal thickness, successful welding is possible.

Fit-up of parts is a constant manufacturing-problem in sheet-metal assemblies and G.S.M.A. allows the welder to cope with this. Gaps between joints may not be constant and a copper backing-bar as shown in figure 4.6b may be used here in a different way. In the open-square butt-joint in figure 4.6c a copper backing is provided with a slot into which a steel strip is laid of the same thickness as the overlapping sheets. The joint can now be satisfactorily welded, the steel strip becoming a part of the weldment. Owing to the comparatively larger volume of metal to be deposited with the gap condition and the quench effect of the copper, a higher wire-speed (current) and output voltage would be necessary than if welding a closed butt with no back-up. To maintain the relative over-all dimensions of the workpiece in this instance, it would be necessary to clamp the whole assembly together.

If a tapered gap is made between two sheets, as shown in figure 4.7a, practical experience in setting the welding conditions and in torch manipulation can be obtained. It will be necessary to clamp the sheets to a copper backing-block. Welding conditions can now be assessed by first making a number of trial welds on closed butt-joints and gradually increasing the gap (without copper backing). Starting the weld on the tapered-gap joint, travel speed will become

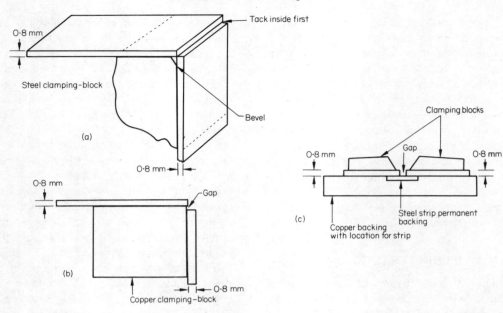

0·8 mm

Tack inside first

Steel clamping-block

Bevel

(a)

0·8 mm

0·8 mm

Gap

(b)

0·8 mm

Copper clamping-block

Clamping blocks

0·8 mm              Gap              0·8 mm

(c)

Steel strip permanent backing

Copper backing with location for strip

*Figure 4.6    Backing bars*

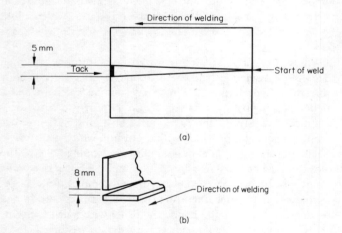

Direction of welding

5 mm

Tack

Start of weld

(a)

8 mm

Direction of welding

(b)

*Figure 4.7    Pre-setting method — 0.8 mm steel sheet*

increasingly slower and lateral torch-weaving increasingly greater.

The electrode extension at the start would be about 10 mm but as the joint becomes wider, the 'stick-out' can be slightly increased (to about 15 mm). The reason for this is to avoid melting back the sheet edges which will be immersed in a relatively high volume of molten weld-metal. Although there is a heat sink in the form of the copper backing, the rate at

which heat can escape to the copper in the region of the fused edges, is to some extent regulated by the much lower thermal-conductivity of the steel.

Figure 4.7b shows a fillet assembly with a tapered gap from 0 to 8 mm. It is not necessary to use a copper back-up for this during practice, but if this was part of an actual assembly, a back-up would be required. When welding this joint from the closed end towards the gap, the electrode is biased towards the horizontal sheet, so that the metal deposited here helps partly to fill the gap. As welding progresses the electrode is moved from the horizontal sheet to the vertical using a slow weaving motion and gradually increasing wire stick-out. When the end of the joint has been reached the stick-out may be as much as 25 mm.

(*Note* This exercise will not produce satisfactory welding, it is only for the purpose of assessing the effect and control achieved by varying the stick-out.)

### Tracking

Greater accuracy in tracking the electrode is required in G.S.M.A. than with fluxed-electrode welding.

To acquire tracking skill metal plate can be prepared as shown in figure 4.8. The lines are best marked by using centre-punch dotting or scribing since chalk lines are not easy to see through a welding filter. Any configuration of lines can

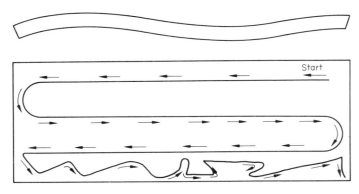

*Figure 4.8    Tracking the electrode*

be used and the plate should be twisted or bent irregularly. This will not only provide good practice in tracking but also in controlling stick-out (or wire extension) beyond the contact tip. Particular attention should be paid to the wrist and arm movements when travelling round the curves.

If a comparison is made here with flux-coated-electrode welding, it will be noticed how much easier it is with G.S.M.A. to execute such intricate manipulation. With flux-coated-electrode welding, slag control would increase the difficulty especially in the vertical or inclined plane, depending on how the plate has been folded. Another advantage of G.S.M.A. can be seen in figure 4.9a, showing the deposition of a weld near the edge of thin material. Using G.S.M.A. it is possible to make a weld right up to the sheet edge without melting it back, but with fluxed-electrode welding this would not be possible — at least the weld width away from the edge would be required on thin sheet.

The assembly in figure 4.9b would again be easy to fabricate with G.S.M.A. and difficult with covered electrodes.

The lap joint in figure 4.9c would also be a problem with covered electrodes, even if the sheets were tightly clamped. The necessity of travelling fast and at the same time manipulating the slag on such light-gauge material would highly tax the skill of the welder. G.S.M.A. welding would be easier but accuracy of tracking would have to be carefully maintained.

Tacking an assembly in sheet metal is much simpler with G.S.M.A. than with covered electrodes or oxy-acetylene techniques. Figure 4.10 shows a series of tacks made along a fillet joint in 1.6 mm material. Although these tack welds were made using a timer to control the wire feed, equally neat and accurately disposed tacks can be made manually with practice.

The wire feed can be stopped and started over a chosen period with a timing device so that the amount of metal deposited can be closely controlled as can the weld length. If the periods of wire feeding are of short duration (say 2 seconds on, 1 second off) a succession of overlapping welds can be made as shown in figure 4.11. Because there is an 'offtime' (a period when the electrode is stationary) between each weld, heat can flow away from the deposit into the surrounding material. This means that a higher rate of heat input, or hotter welding-condition, can be used, thus increasing weld-metal fluidity.

Some typical joint configurations used in light-gauge assembly are shown in figure 4.12. The joints in this range, 0.6 to 3 mm thick, seldom require edge preparation. The closed butt (figure 4.12a) is ideal for G.S.M.A. welding and especially for automatic application. The open butt (figure 4.12b) can also be machine welded but would require backing. The tee

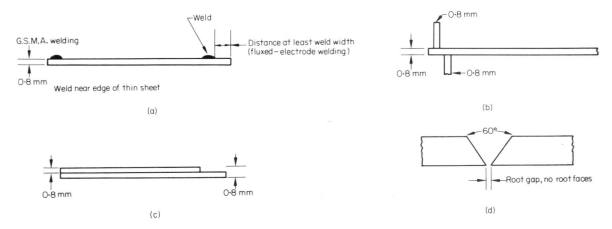

*Figure 4.9    G.S.M.A.  compared  with  flux-coated-electrode welding*

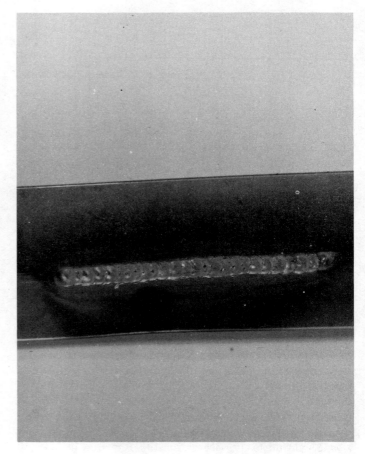

*Figure 4.10    G.S.M.A. timed tack-welds*          *Figure 4.11    Pulsed electrode G.S.M.A. weld*

joint (figure 4.12c) is easily welded in any position and a massive-section copper-backing at the rear of the horizontal member will minimise distortion.

The lap joint (figure 4.12d) is best assembled by means of 'spot' or resistance welding before fusion welding the edge. In welding a lap joint the greater volume of deposited metal is made on the lower sheet. The edge of the advancing weldpool should then exactly coincide with the edge of the top member. If the arc is allowed to wander too far across this line the top sheet will melt and run back from the weldpool. Let us consider why this may happen. The weldpool is mainly deposited on the lower sheet which is also exposed directly to the arc heat. Only the side or edge of the top sheet contacts the molten metal therefore most of the heat spreads away

from the weld area through the bottom sheet. The comparatively limited area of the top-sheet edge does not therefore transfer heat away at the same rate.

If the lap is open as in figure 4.12e the difficulty of welding will be greatly increased, because in addition to the reasons given above, the top sheet is not in contact with the bottom sheet in the weldpool area. This means that heat cannot escape into the weldpool area, and therefore the risk of melting back the top sheet is intensified.

Close tacking at about 25 mm pitch will help to control and maintain the original gap. Without tacking, the gap, especially on a long joint, would progressively increase with the advancement of the weldpool. Increasing the electrode stick-out will provide better control while welding this assembly.

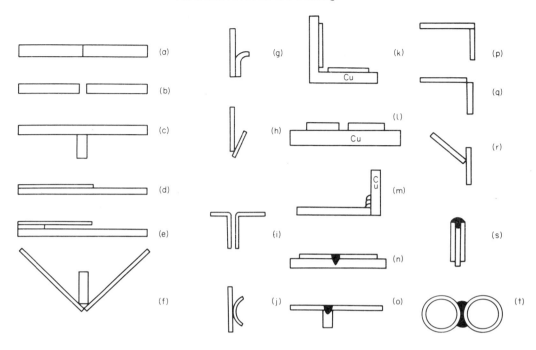

Figure 4.12    Joint assembly and comparison

The joint in figure 4.12f would be difficult to weld from the inside because of limited access; a fully penetrated weld could, however, be made from the outside, joining the three members.

The configuration in figure 4.12g would not be a good preparation for fluxed-electrode welding because it would tend to trap slag at the root and produce undercut on the radius. G.S.M.A. welding does not suffer from either of these disadvantages and would therefore be satisfactory.

The joint in figure 4.12h is also suitable for G.S.M.A. welded along the outside corners.

The preparation in figure 4.12i presents an inevitable slag-trap because of the gap between the interfaces of the joint. G.S.M A. would again be a satisfactory welding process. The joint in figure 4.12j would produce slag trapping and severe undercutting if fluxed electrodes were used. Even G.S.M.A. welding may produce undercutting here because of the tendency to wash down metal from the radii into the grooves.

Figures 4.12k and l illustrate the application of heat sinks in the form of copper backing-bars and in figure 4.12m a vertically situated copper block is used to align successive

weld-beads to form a section at right angles to the base or parent material. Plug, or G.S.M.A. 'spot'-welding applications are shown in figures 4.12n and o. The corner joint in figure 4.12p is ideal for G.S.M.A. welding but tacking should, if possible, be confined to the inside to keep the outer weld smooth and consistent. The corner joint in figure 4.12q is better because it gives a full fillet with equal leg length and root penetration. The joint shown in figure 4.12r would require close pitch tacking before welding otherwise severe buckling would cause the oblique member to move away from the vertical one, or vice versa, depending where the arc and weldpool are directed preferentially.

The three-sheet assembly (figure 4.12s) is satisfactory for welding but easier in the flat or vertically down positions; horizontal or overhead welding would be difficult. The tube sections in figure 4.12t would present few problems to the G.S.M.A. welder but would be extremely difficult to accomplish without melt-through or undercutting by fluxed-electrode welding. This applies only to light-gauge metal; in rolled plate or thick-walled tubes (say, 8 mm thick) the problem would not be so acute but the inherent risk of slag trapping would still be there with metal-arc welding.

## Vertical Welding

If a 20 mm single-pass vertical fillet-weld were to be made with fluxed electrodes considerable skill would be required of the welder. Although G.S.M.A. does call for skill in manipulation, there is no slag to contend with and therefore positional welding is that much easier. In light-gauge material the weld can be made vertically downwards in one or more passes. The angle of the electrode to the weldpool is determined by the necessity to control the molten metal. If, in a vertically down weld, the weldpool is allowed to run in advance of the electrode, a cold lap or cold shut may occur. This is an area where there is an absence of weld-metal fusion into the joint. It constitutes a serious welding-fault that is all the more so because it may go undetected.

Vertically upwards welding demands more skill, more attention to electrode manipulation and a great deal of concentration. To achieve the necessary skill in this field, intensive practice with flux-coated electrodes should be undertaken; once the welder is adept in this field, G.S.M.A. welding will prove comparatively easy.

### Overhead

A short arc, without weaving, is used wherever possible and attention must be paid to correct tuning of the inductor so that transfer of metal from electrode to weldpool is even and without explosive short-circuits.

### Horizontal—Vertical

A succession of overlapping runs are used for depositing large vee butt- or fillet-welds, each bead half overlapping the previous one, as in flux-coated electrode welding.

Some typical settings for positional welding would be as follows.

*Vertically down:* butt in 3 mm plate using 0.8 mm wire (single pass), 120 A, 200 V.

*Vertically up:* butt in 0.5 mm plate joint single-vee with no root gap or root face using 1.2 mm wire (3 passes), 160 A, 20 V. Fillet welds would require the same conditions: over-head fillet in 0.50 mm plate using 1.2 mm wire (3 passes), 140 A, 20 V.

In the welding of thick-walled pipes it is customary to put the first run in with T.I.G. ('argon arc') and follow with successive runs of either G.S.M.A. or flux-covered electrodes (as described on p. 41, the stovepipe technique).

Since pipe joints are ideal for automatic welding this is most often employed in workshop fabrication with manual welding used mostly *in situ*. Automatic welding is, however, becoming increasingly popular for field work. The risk of defects due to faulty operator-technique is minimised by machine welding. Pipe and boiler work are specialised fields of welding requiring highly skilled welders in all of the fusion techniques.

For details of joint design reference may be made to *Recommended Welded Connections for Pipework*, published by the Institute of Welding (1965).

The range of application of G.S.M.A. welding extends from about 0.6 to about 70 mm thickness material although the factor known as the weld-depth-to-width ratio becomes critical in the thicker material. Cracking and weld failure are therefore more likely and to combat this special care is needed in calculating joint geometry; for example, a wide inclusive-angle of about 70° for a double-vee butt-preparation in, say, 50 mm plate would be required, or a double 'U' with no root-gap and a root radius could be used, as shown in figure 4.13.

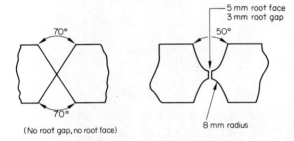

*Figure 4.13    Some joint preparations for heavy plate*

Of the two examples, the double-vee butt is the least costly to prepare because it can be produced by flamecutting, while the double 'U' joint would require relatively expensive machining. In some situations, for example, where very large plates are to be prepared, flamecutting would provide the most practical preparation-method, the edges being either cut on a large pantograph or by using a portable machine running along a track. Arc air gouging can be used for obtaining a vee profile, but this method, although now available in mechanised form, is not as widely employed as flamecutting.

## Alloy Steels

When welding medium-tensile low-alloy steels in light gauge with the G.S.M.A. process, an approximate matching of parent-metal composition is achieved in the weld metal by the effect of dilution. In heavy-gauge alloy steels this mixing of the weld metal with the parent is only significant in the first few runs of a multi-pass deposit. Therefore to produce a joint of strength compatible with the material that is welded, extra alloying elements are incorporated in the wire; these include those with a high affinity for oxygen (deoxidisers) like aluminium, titanium, silicon and manganese.

Elements like chromium, nickel and molybdenum are added to confer special properties. For example, chromium and nickel are the essential alloying elements in the various stainless steels. Both chromium and molybdenum confer better hardening-properties on steel, and together with vanadium and carbon are used for tool and die steels, many of which require special welding-techniques, pre-heat and post-heat, in order to develop and maintain their physical properties. Cast steel can be welded with G.S.M.A. but cast iron can only be welded using special techniques.

## PULSE-ARC WELDING

Pulse-arc welding is a derivative of G.S.M.A. and offers increased scope for welding by the consumable-bare-wire technique of a large number of non-ferrous and ferrous materials. It was developed for the welding of aluminium, and requires the use of two power-sources. A selected high-current pulse is added to a low-current established arc giving an average current-level suitable for the wire speed. The high-current pulse provides droplet detachment of the globule forming on the electrode tip and with an argon-rich gas-shield the drop has an axial path. (That is, the droplet falls perpendicularly, unlike the condition that would exist if carbon dioxide were used, when the globule would be transferred to the weldpool by deflection due to cathode force.)

For welding low-alloy steel a gas mixture consisting of 95 per cent argon + 5 per cent carbon dioxide is used. Stainless steel can also be welded using this gas although for metallurgical reasons, 99 per cent argon + 1 per cent oxygen is often used.

Standard G.S.M.A. equipment can be used for welding light-gauge aluminium alloys using matching filler-wire and pure-argon shielding. Wire diameters of the order of 0.6 to 0.8 mm cannot, however, be pushed over any great distance because of their softness. One way to overcome this problem is to use a welding gun that contains a miniature reel and wire feedrolls with a small drive-motor. However, this unit is cumbersome and difficult to maintain accurately over the joint line. Another system, called 'push pull', uses the standard 13 kg reel of wire pushed in the normal way but assisted by a small pulling-motor in the welding gun.

Pulse-arc welding does not require the use of fine wires; for thin aluminium, wire of 1.6 mm diameter, for instance, can be used to weld a butt joint in 2 mm plate. Thick wire does not present the difficulty in feeding and it is proportionately cheaper.

The two main parameters of pulse-arc welding are *pulse frequency* and *pulse amplitude*, the latter being determined by the voltage. Incorrect voltage-adjustment will result in an unstable arc and metal-transfer conditions. Pulse-arc welding offers precise penetration-control and can be used effectively in all welding positions. Vertical welding, however, as with all the fusion processes, demands greater manipulative skill. Butt welds in material as thin as 2 mm can be made with 1.6 mm wire, but with increasing thickness joint preparation is required. A single-vee preparation for 25 mm plate, for example, would mean an 80° inclusive angle and a root face of about 3 mm with no root-gap. A multi-run deposit would be necessary, taking care to achieve satisfactory root-penetration with the first run. The welding of aluminium pipes and vessels *in situ* calls for the same degree of operator skill demanded for metal-arc welding, except that there is no slag-control problem.

Automatic welding is substantially better for pipe work and is used for high production where the workpiece can be rotated. In the field there are now available automatic welding-heads that travel round large-diameter pipes using small rack-driven tractors.

For manually welding thin-walled pipes, pulse-arc welding has the great advantage of not requiring copper or steel backing.

Heat-resisting medium-carbon steel (using 90 per cent argon + 5 per cent carbon dioxide), stainless steels, nickel and its alloys, copper and bronzes (using 99 per cent argon + 1 per cent oxygen) can also be pulse-arc welded.

## FURTHER STUDY

A. A. Smith, *Carbon Dioxide Shielded Consumable Electrode Arc Welding*, Welding Institute, Cambridge, 1965.

The Welding Institute, *Recommended Welded Connections for Pipework*, Cambridge, 1965

K. P. Bentley, 'Element transfer in $CO_2$ welding', *Br. Weld. Res. Ass. Bull.*, 7, 204–9, 1966

K. W. Brown, 'Programmed M.I.G. welding', *Metal Constrn and Brit. Weld. J.*, 1, 286–90, 1969

# 5. GAS-SHIELDED TUNGSTEN-ARC WELDING

Gas-shielded tungsten-arc welding (called 'T.I.G.', 'tungsten—inert-gas', or 'Argon-arc') is a fusion process that uses a virtually non-consumable electrode in a completely inert gas shield. The metal tungsten has the highest melting point of all metals, at 3410 °C. (Although it is often called non-consumable, this is not strictly true since it does gradually erode away in the arc.) The only alternative material is carbon which, although it does not melt, is dissipated comparatively rapidly in an arc, even if an inert-gas shield is used. In addition, carbon electrodes of small diameter are extremely fragile.

Figure 5.1 shows the elements of G.S.T.A. welding. The power source can be a.c. or d.c., depending on the metal to be welded. In this form of welding it is not good practice to short-circuit the electrode to the workpiece as in G.S.M.A. or metal-arc welding. This is because there are inclusions of tungsten in the workpiece which may cause cracking, and also because by short-circuiting the electrode it may become chipped or contaminated. In order to strike an arc, a superimposed high-frequency discharge is used. This is supplied from a special high-voltage generator and has a frequency of about 3 MHz. Since this is a broadcasting frequency care must be taken in some locations to screen the source to prevent radio interference.

The current regulator (unlike, say, the one used with metal-arc welding) must be capable of fine adjustment because even 1 A variation may be significant when welding very thin materials.

For manual operation, the high-frequency initiating-spark and welding-current are controlled by a foot-operated switch. This often has two contacts so that the high-frequency discharge precedes the flow of welding current.

Because of the high temperature developed at the arc and the subsequent heating of the electrode, water cooling is needed for the torch. (There are exceptions, for example, when a miniature torch with a maximum-operating current of, say, 70 A is used.) Special material that can withstand

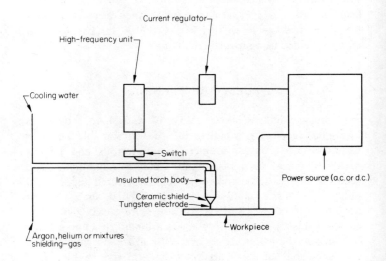

*Figure 5.1    G.S.T.A. system*

prolonged heating and provide positive insulation for the high-frequency current is used for the torch body. The gas shroud may be made of ceramic or consist of a water-cooled copper vessel.

Argon is most often used as a gas shield in the United Kingdom but in the United States helium or helium—argon mixtures are common. Argon is a rare gas obtained by the fractional distillation of liquid air; it constitutes only 1.3 per cent of the atmosphere, is monatomic and completely inert. The element helium does not occur naturally in the United Kingdom or Europe, the United States being the world's major source where it is obtained from oilwells. Helium has a density ten times less than argon; because of this the flow rate can be two to five times higher than argon. Helium gives increased penetration, higher travel-speeds and requires less current than argon. The heli-arc is about twice as hot as the argon arc but

*Figure 5.2    G.S.T.A. welding* in situ *(courtesy of B.O.C. Ltd)*

requires very high initiating-voltages and is therefore more suitable for automatic application or for welding thick material.

Using the G.S.T.A. process a wide variety of metals and alloys that had been difficult or impractical to weld by other techniques, could now be welded. Among these are magnesium, aluminium, copper and their alloys, silver, stainless steel, nickel and special heat-resisting alloys. Figure 5.2 shows aluminium being welded; note the relative angles of the torch and filler wire — these are most important in G.S.T.A. welding.

## ELECTRODE GEOMETRY

Every craftsman, whatever his medium, must constantly repair, sharpen and take good care of the tools of his trade. The worker in wood, for example, must constantly maintain the cutting edges of his tools; the welding craftsman is no less responsible for keeping his blowpipes, nozzles, contact tips electrodes, etc., in good working order.

The tungsten electrode, being semi-permanent, requires special attention. The size and end shape affect durability and

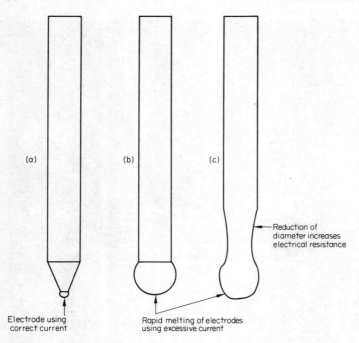

Figure 5.3  *G.S.T.A.  electrode  characteristics  (negative polarity)*

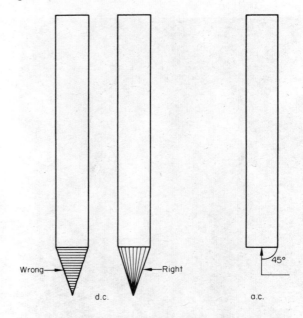

*Figure 5.4    Electrode grinding*

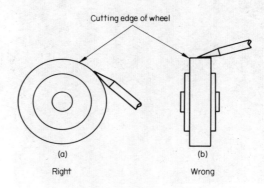

*Figure 5.5    Use of grinding wheel*

such things as weld throat width and penetration. Other factors related to electrode geometry are the voltage—current relationship of the arc and the temperature gradient along the electrode. Using high currents, for example, the tip of the electrode tends to be cooler than the immediately preceding portion. This phenomenon is accounted for by the thermal energy loss due to electron emission at the tip. In figure 5.3a the condition of the electrode after use with normal current and negative polarity is seen. In figure 5.3b the whole of the cone end has been melted by using a current greater than the electrode's conductive capacity (this effect is also produced more easily with electrode positive). In figure 5.3c the current used was so excessive that in addition to the balling of the tip, necking has occurred.

For d.c. supply electrodes should be ground with a conical end and for a.c. with a flat end as shown in figure 5.4. The way in which the electrode is ground affects its performance; this applies especially to the fabrication of light-gauge materials where very low currents are used. For producing the correct cone shape the electrode should be held as in figure 5.5a. The side of a grinding wheel should not be used because grooving will occur and this will weaken the wheel making it unsafe. If the cutting edge becomes worn and irregular, there

are tools available for restoring the correct contour. By grinding the electrode as in figure 5.5b, a series of nodes or projections are produced which, by their relative orientation to the longitudinal axis of the electrode provide arc instability points. Thus as the electrode becomes hotter, so the tendency for the arc to climb away from the tip increases. If the grinding marks follow the line of the electrode longitudinal-axis, arc climb is considerably reduced.

Other factors that promote unstable arc conditions due to electrode malformations are shown in figure 5.6. Perhaps the most common condition is shown in (a). During welding, a

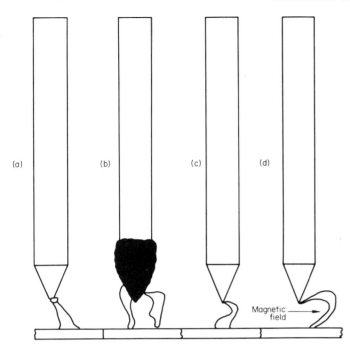

Figure 5.6    Electrode faults

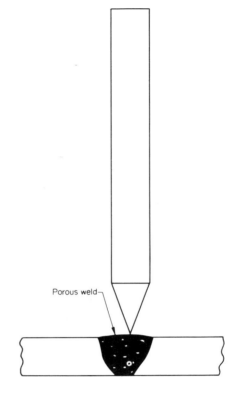

*Figure 5.7    Porosity — common causes are: electrode contaminated, work contaminated, magnetic fields (arc blow), insufficient current, insufficient shielding-gas, contaminated shielding-gas, wrong filler material, incorrect torch manipulation, poor current-return path, side draughts, electrode melting, excessive electrode stick-out, arc to long, electrode shorted to work*

tiny globule of weld metal may be thrown out of the weldpool and stick to the tip. Deflection of the arc will follow because a new node or projection has been formed on the electrode. The condition may be further aggravated if the electrode is allowed to cool after welding in insufficient argon. The resulting oxidation of the electrode will mean arc striking will be difficult and when the arc is finally initiated it will tend to become established on the least oxidised portion. Although the contaminating layer may subsequently be sputtered off the tip, if oxidation is not too severe, this is not good practice. In figure 5.6b where gross contamination of the electrode is shown, the only remedy is to regrind the tip completely, making sure that all the contaminating material is removed. Such a condition is common with beginners and happens for three main reasons

    1.  immersion of the electrode into the weldpool
    2.  short-circuiting of the electrode to the work, even though no molten pool is present
    3.  contact between the filler wire and the electrode.

Damage to the electrode may also produce unstable welding conditions as shown in figure 5.6c. When using low-current arcs that only have a feeble column-strength, quite small

magnetic fields may promote arc wander as in figure 5.6d. This applies especially when using d.c. and in such conditions as the welding of edge, inside corner and butt joints in thin material.

    Wrong electrode-shape may also be a cause of weld porosity by the fact that unstable-arc conditions may be promoted — see figure 5.7. Other common causes of porosity are suggested but these by no means exhaust the list. Porosity may, for example, be due to intermetallic reactions or the release of existing trapped gas in the workpiece.

    The tip-angle preparation for use with d.c. should be about 60°; if it is more, say, 45°, the arc will tend to wander from the point. For a.c. welding a flat smooth-faced electrode and preparation are required. For best results both types should be machine ground.

## SAFETY

The usual precautions applying to the operation and use of electrical apparatus apply to G.S.T.A. welding, with some additions. The intense arc plasma generated with a tungsten electrode in argon evolves not only more ultraviolet radiation than metal arc or G.S.M.A. but under some circumstances generates a large amount of ozone, $O_3$. This gas, which is chemically very active, can be a hazard to personnel, particularly if a number of G.S.T.A. welders are working in close proximity with inadequate fume-extraction. Ozone is generated along with oxides of nitrogen (also harmful) when an electrical discharge passes through the air, (as in a thunderstorm, for example) and has a characteristic odour — acrid, acid or metallic. The oxides of nitrogen are not so recognisable.

The high-frequency alternating current used for arc ignition requires high-voltage insulation. Standard low-voltage insulation should not be used. Because the h.f. unit also generates radio frequencies care must be taken to screen the source effectively to avoid interference with local radio-receivers.

If rotary generators are used, capacitors of the correct value must be fitted across the output leads to prevent insulation breakdown of the motor windings. High-frequency discharges can cause skin damage by contact — avoid touching electrodes with your hand, even if gloves are worn. Make sure no other electrical apparatus is in contact with the welding bench, the workpiece or any part of the welding circuit — damage to insulation may occur if this happens.

Highly polished surfaces of metals like aluminium and stainliss steel reflect considerable arc-radiation — make sure that the appropriate shade of welding glass (related to current used) is fitted in the screen. Avoid wearing white overalls — these reflect infrared and ultraviolet radiation.

Do not wear open-neck shirts, or work with bare arms; if you do severe local skin burns may result.

Never handle any part of the apparatus with wet hands or wet gloves.

If someone is standing near remember to give warning before you strike an arc. Never use any other gases but argon, helium, or mixtures of these as directed.

Do not attempt to adjust the spark gap inside the high-frequency unit, as much as 5000 V may be generated at this point.

## THE MAIN PARAMETERS OF G.S.T.A. WELDING

These are: (1) arc voltage, (2) current, (3) travel speed, (4) arc-initiation time (especially in automatic welding), (5) gas-shield

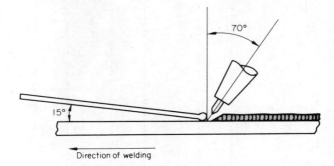

*Figure 5.8    G.S.T.A. welding technique*

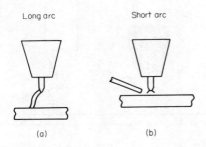

*Figure 5.9    Arc length; (a) long arc causes arc instability, porosity and inconsistent penetration; (b) short arc—arc gap just wide enough to allow unrestricted insertion of filler wire*

flow-rate and type, (6) electrode shape, (7) arc gap, (8) electrode attitude to workpiece, (9) gas purge and gas delay.

For depositing a weld on a horizontal surface (figure 5.8) the electrode should be held at 70° from the vertical, as shown. The filler-wire angle must be shallow, about 15° from the horizontal, to give the operator good weldpool visibility. The procedure is similar to oxy-acetylene welding with the difference that the distance from the heat source (that is, the electrode tip) is much more critical than with the cone of an oxy-acetylene flame. Just sufficient clearance between the electrode tip and the weldpool must be left to allow access of the filler rod. The arc, besides being kept as short as possible, must not vary in length while traversing the joint. This applies especially to the welding of light-gauge materials with a low-current arc. An excessively long arc in any situation can lead to weld-metal contamination, arc instability and porosity. The correct arc-length is that which just allows access of the filler wire to the weldpool (see figure 5.9). An excessive amount of shielding gas is not only wasteful but may cause weld porosity by air entrainment. Jets of gas issuing from small-diameter holes into a surrounding tube will, if the tube is

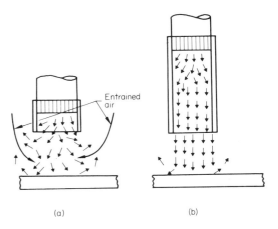

*Figure 5.10    Gas flow; (a) turbulent flow, with entrained air, (b) laminar flow*

relatively short, produce turbulence at the point of issue (see figure 5.10a). Turbulent flow causes induction of air. If the tube length is correct, the issuing gas will be stiff or columnar, this is called laminar flow (figure 5.10b). Commercial torches are designed to operate at their recommended gas pressures related to torch orifice and shroud geometry; if gas flow is greatly exceeded, conditions of laminar flow are upset and turbulence will result. Torches are also assessed with a maximum-ampere rating at a stated duty cycle. This should not be exceeded. Water-flow fuses are often employed, but care must be taken to ensure that the minimum amount of water is flowing according to the advised rating. If water is recirculated through a heat-exchanger system the maximum-current capacity of the torch should not be exceeded even for a short period. Thoriated or zirconated electrodes should be used when available, since they require slightly less current, besides giving more stable arc-conditions. If ceramic shrouds become grossly overheated, improvements may be made by using a shroud of larger diameter or reducing electrode diameter and current.

## Electrode Attitude and Tracking

The arc in G.S.T.A. welding cannot be varied as much as in metal-arc welding. When welding very thin material (1 mm thick, say) the arc length may be only 2 mm long. Extension of this to 4 mm (without altering other parameters such as current, travel speed, etc.) would cause a variation in weld width and penetration.

When using this process for welding thin material, the welder must cultivate a very sure and steady hand and have an unrestricted view of the joint line. Since the electrode may only be 1.2 mm in diameter and pointed, it requires great concentration to maintain torch attitude, arc gap and accurate tracking of the joint. This applies especially to tight butt-joints in polished metal where reflected light adds to the difficulty of joint-line observation.

For autogenous welding (using only the parent material to obtain fusion) of thin sheets using a square butt preparation, two hands are better than one for tracking (see figure 5.11). The plates to be joined should be tightly aligned by clamps (these are not shown in the diagram). The workpiece edge A should be at eye-level. Welding should be carried out sitting down — supporting the body weight and allowing the arms to accommodate the necessary torch-movement better. The elbows should rest against the bench surface so that the total weight to be sustained is only from elbows to finger tips.

As the torch is drawn along the joint towards the body of the operator, a perfect view of the arc, electrode, weldpool and torch angle is obtained. Fine angular-adjustment of the torch is made by the finger and thumb of the left hand. Since motion along the joint is towards the body, the wrist movement is inwards, describing a natural arc. Using this attitude, very precise control of penetration and weld width can be obtained.

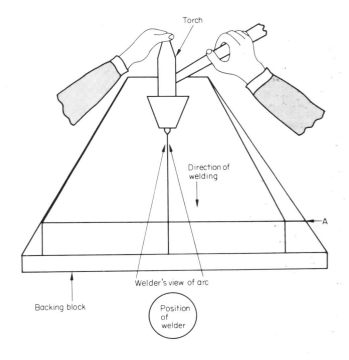

*Figure 5.11    Welder-to-work relationship*

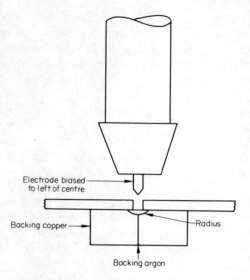

Figure 5.12    *Backing with argon*

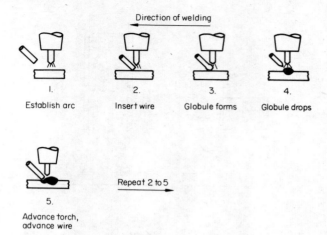

*Figure 5.13    G.S.T.A. procedure*

The importance of locating the electrode tip exactly on the centre line of a butt joint must again be strongly emphasised. In figure 5.12, the electrode is shown off-centre to the left over an open-square butt-preparation. This attitude will promote gross melting away of the left-hand sheet, and in the case of an automatic or machine-tracked torch, the situation would not be so readily retrievable as with manual welding. In such a situation the manual operator would adjust his torch-to-work relationship, unlike the machine, which cannot think!

### Filler Wire

The mode of applying filler wire to a joint using G.S.T.A. is very similar to that used in oxy-acetylene welding. The main difference lies in the greater accuracy and orientation of the filler wire required in G.S.T.A. welding. Figure 5.13 shows the sequence. The application of the wire must be precise and regular so that the globule reforming on the wire tip is just the right size, and is exactly over the crater at each forward movement of the wire.

### JIGGING AND JOINT PREPARATION

The butt, fillet, lap, edge and corner joints in light-gauge metal are ideal for G.S.T.A. welding. The parts to be welded should be held in precise alignment using bolts, straps and screw clamps, or, for machine welding, hydraulic pressure. The

intense concentrated heat produced by the tungsten arc means that high travel-speeds can be reached if work fit-up is correct. For closed butt-joints a backing bar, although not essential, greatly assists control of penetration. The significance of a root gap in relation to the molten weld-metal can be seen in figure 5.14.

The crucible, or ladle, as used in a foundry, is a steel vessel lined with refractory materials like silica. Since silica has a melting point considerably above that of molten steel and is also a relatively poor thermal-conductor, the ladle wall is protected.

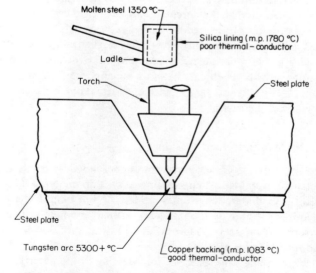

Figure 5.14    *Thermal characteristics of materials*

Consider now the single-vee joint. This has a root face and a root gap. If there were no root gap the joint would perform a similar function to a ladle in that it would contain molten metal. However, since the workpiece is of the same material as the deposited weld-metal, the ability of the joint to contain liquid is limited. Travel speed, electrode presentation, current, work position, skill (or lack of it) in handling the torch, etc., may each produce burn-through or collapse of the root faces of the joint. To offset this possibility a false bottom can be clamped to the 'ladle'/vee joint. A copper bar beneath the joint will allow metal to run over its exposed surface and solidify without fusing to it. This can be accounted for not because the melting point of copper is higher than that of steel, (in fact copper melts 267 °C lower than steel) but because copper is an excellent thermal-conductor. The weld metal is therefore cast on the copper and the underside of the welded joint requires very little finishing. However, care must be taken when using copper backing that the section is massive enough to sustain the work cycle; thus a copper section of 2 mm thickness used for backing single-vee joints in 25 mm thick steel would melt owing to thermal overloading. For manual welding a massive copper section is usually adequate, but for machine welding, and especially when the metal to be welded also has a high thermal-conductivity, like aluminium, water-cooled copper bars are necessary.

In addition to good jigging, the working position relative to the welder should be considered. For small sections or short weld-runs, the workpiece should be positioned so that the easiest natural wrist-movements (right hand, working towards the body — leftwards) can be executed, as shown in figure 5.15. Because of the extremely fine adjustment necessary for welding thin material with a light-weight torch it should be held lightly, as you would hold a pencil. So the forefinger and thumb are used for delicate manipulation and the wrist is used for traversing the joint. Working in this manner, the body weight should be supported by sitting, leaning, or standing with your feet apart.

Before starting to weld, the correct body-adjustment should be made and the torch (with current and h.f. switched off) tracked along the joint line. Sometimes it will be found that by doing this the finishing of the joint is going to be difficult because of incorrect workpiece relationship to the body, or perhaps a jig member may be in the way. If so, start again and repeat the exercise until a clear run, *within the extension capability of the arm and hand holding the torch*, is assured.

To execute straight lines, a sign writer uses a 'steady'; this is simply a straight edge which acts as a reference line for his brush. In G.S.T.A. welding the same principle can often be

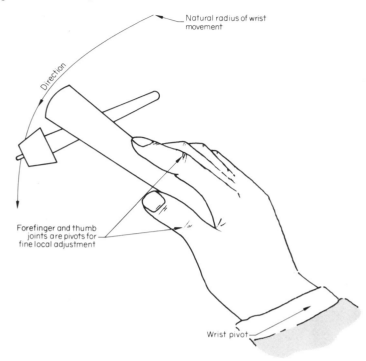

*Figure 5.15     Hand and wrist motion*

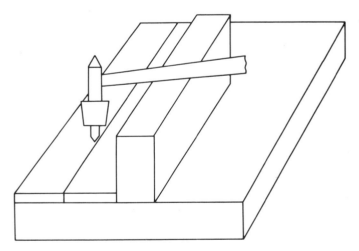

*Figure 5.16     Using guide bar*

applied, as shown in figure 5.16. In practice, the third finger of the right hand acts as the 'sensor'. Keeping the tip of the finger in contact with the guide-block surface while tracking the torch along the joint is easier than keeping the torch in direct contact with the guide. This is because if the torch is pressed against the guide block, it may produce a jerky intermittent motion because of friction, while the sensitive finger-tip will adjust the bearing pressure accordingly.

## WELDING ALUMINIUM AND ITS ALLOYS

Metals like aluminium, magnesium and their alloys, which form high-melting-point refractory surface-oxides, are usually welded using an a.c. power source. The metal to be welded largely determines whether a.c. or d.c. is to be used — stainless steel, titanium and copper, for example, are best welded with a d.c. supply. Table 5.1 lists welding conditions for aluminium and its alloys.

*Table 5.1    Welding conditions for the a.c. welding of aluminium and alloys*

| Material thickness (mm) | Current (A) | Electrode diameter (mm) | Argon flow (litres/min) |
|---|---|---|---|
| 0.55 | 25—30 | 1.6 | 4.0 |
| 0.7 | 35—40 | 1.6 | 4.5 |
| 0.9 | 45—60 | 1.6 | 5.0 |
| 1.2 | 60—70 | 2.3 | 5.0 |
| 1.6 | 75—85 | 2.3 | 5.0 |
| 2.0 | 95—100 | 3.3 | 6—7 |
| 2.5 | 100—120 | 3.3 | 7.0 |
| 3.3 | 130—145 | 3.3 | 7.0 |
| 4.9 | 145—190 | 3.3 | 7.5 |
| 6.4 | 210—250 | 5.0 | 8—9 |
| 10.0 | 260—320 | 5.0 | 9—12 |
| 12.7 | 330—390 | 6.4 | 12—15 |

(Filler wires range from 1.6 to 6 mm)

With d.c., about two-thirds of the arc heat is generated at the positive pole. If the electrode is positive the tendency will be for it to melt rapidly because of overheating. Conversely, with the electrode negative, the workpiece receives a greater proportion of the heat and the electrode is relatively cool.

When welding aluminium with an a.c. power supply a condition known as inherent rectification occurs. In a 50 Hz supply, voltage/current reversals occur 100 times per second, and the current flowing during the negative half-cycles is greater than that flowing during the positive half-cycles. To prevent damage to the equipment and safeguard the operator it is therefore usual to incorporate a large bank of electrolytic capacitors into the welding circuit.

The refractory surface-oxides present on aluminium are sputtered off by the action of the a.c. arc (see figure 5.17) and the weld deposit is, under correct welding conditions, clean and uncontaminated.

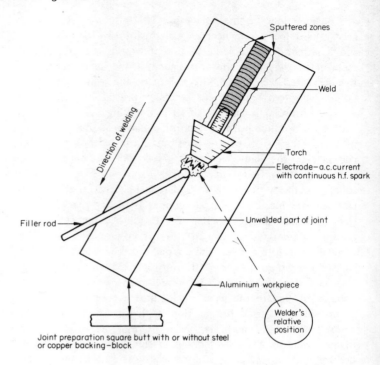

*Figure 5.17    Cathode sputtering*

Argon purity is a vital factor and only the highest grade should be used. No dilution by any other gas (except for the welding of very thick material with helium—argon mixtures) should be made. Another function of the shielding gas is to protect the heated metal ahead of the arc from progressive oxidation (see figure 5.18). Because aluminium is such a good thermal conductor, heat from the weldpool travels ahead and so the rate of surface-oxide increase with temperature is proportionally rapid. Because of the argon flowing along the joint ahead of the arc, protection of the welding area is achieved. Removal of surface oxide before welding is often necessary, even with bright and apparently clean material. Wire brushing or chemical cleaning are often used and it is important to perform the cleaning operation immediately before welding. The prepared surface should not be touched with the fingers afterwards, otherwise contamination will reoccur.

Generally with G.S.T.A. welding torch oscillation is not desirable or necessary except when welding heavy-gauge material. Even then lateral (across the joint) or elliptical torch-manipulation should be made with discretion to avoid disruption of the argon shield and possible air entrainment. Butt joints in thin material, with or without matching filler,

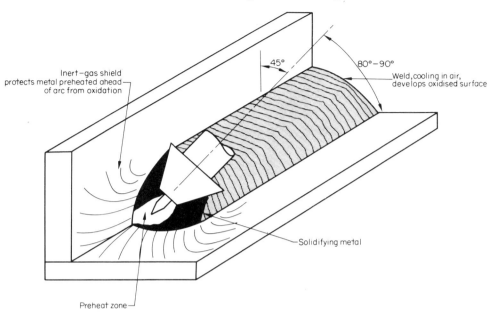

*Figure 5.18    Inert-gas function*

should be made without sidewards movement, the torch being kept accurately over the centre line and moved along it at a constant speed until within a short distance (say about 20 mm) of the end of the weld. Here it is necessary to accelerate because this part of the joint will have received substantial pre-heat from the weldpool.

In fillet welding the torch axis relative to the horizontal plane should be between 80 and 90°, and 45° relative to the vertical wall of the workpiece. Filler material must be supplied to the weldpool at an acute angle, about 10° from the horizontal. The filler metal should be supplied at a constant feed-rate and the molten end kept within the argon shield for the whole welding-time. In fillet welding, the electrode can be extended beyond the gas shroud to increase weld-area visibility. A slight increase in argon flow may be required in this instance.

A gas-purge or pre-flow switch is often incorporated in the equipment. The gas-purge flow-time can usually be varied as can the gas delay provided at the end of the weld. Gas pre-flow ensures electrode protection while gas delay additionally allows the weld crater to cool in an inert atmosphere. It is particularly important to ensure that the gas-delay period, after the arc is extinguished, is long enough to allow the hot electrode to cool below its oxidation temperature.

A sign of inadequate electrode protection is the blue, yellow and purple discolouration characteristic of tungsten cooling in air.

Positional welding of aluminium requires a high degree of manipulating skill, preferably with the workpiece in the horizontal plane, as applies to other fusion-welding techniques.

Filler wire should be cleaned immediately before welding with steel wool because oxides and drawing compounds may be present. It is important to use only the appropriate filler-wire for the job, since there are a great many aluminium alloys with widely different physical-properties.

### Direct-current Welding

For heavy-section aluminium, direct-current welding with electrode negative is also used. The technique is used for automatic applications using a helium or helium—argon gas-shield. Magnesium and its alloys can also be welded using a.c. or d.c. power.

### WELDING STAINLESS STEEL

Stainless steel can readily be welded in any position and may be welded to some other metals. The most common stainless steel is the 18/8 chromium—nickel type. Table 5.2 lists welding conditions for stainless steel.

*Fundamentals of Welding Skills*

*Table 5.2    Welding conditions for the d.c. welding of steels*

| Material thickness (mm) | Current (A) | Electrode diameter (mm) | Argon flow (litres/min) |
|---|---|---|---|
| 0.55 | 15—30 | 1.2 | 1.0 |
| 0.7 | 20—40 | 1.2 | 1.0 |
| 0.9 | 30—50 | 1.6 | 1—2 |
| 1.2 | 40—60 | 1.6 | 1—2 |
| 1.6 | 60—80 | 2.3 | 2.0 |
| 2.0 | 70—90 | 2.3 | 2.0 |
| 2.5 | 90—110 | 2.3 | 3—4 |
| 3.3 | 110—125 | 3.3 | 4.0 |
| 4.9 | 130—165 | 3.3 | 4.0 |
| 6.4 | 170—200 | 3.3 | 5.0 |
| 10 | 225—290 | 5.0 | 5—6 |
| 12 | 300—350 | 5.0 | 6.0 |

(Filler wires range from 0.6 to 6 mm)

Since steels are relatively poor thermal-conductors, care must be taken in the welding of thin sections to avoid excessive distortion. Power clamping of the work gives the best results. Although a.c. can be used for thin sections, d.c. power gives superior welding in all thicknesses.

As with G.S.T.A. processes generally, the closed-square butt-preparation, with or without added filler, is ideal for light

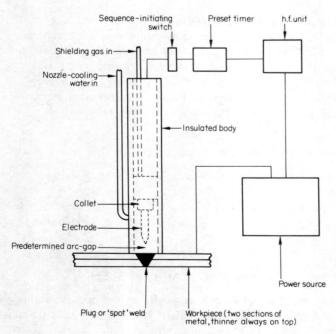

*Figure 5.19    Automatic plug-welding (spot welding)*

material. Distortion with other types of joint (for example, open butt, fillet, vee groove, etc.) is increased because of the added filler and extra heat-input. Copper or aluminium bronze should be used for clamping stainless-steel sections. If steel is used, magnetic fields around the clamp toes next to the joint line may create arc-blow.

Copper may be used as joint backing, but for heavy material or automatic welding water cooling should be incorporated. Workpiece material and filler wire need the same amount of cleaning as other metals, since there is no fluxing action.

Because of the great affinity of chromium for carbon, filler rods are stabilised by the addition of titanium or niobium, since these metals have an even greater attraction than chromium for carbon. The presence of a stabilising additive in stainless-steel plate or filler material is denoted by the label 18/8/1, the 1 per cent being the added element.

If chromium is allowed to form its carbide during welding, there will be a deficiency of the element in the fusion zone giving rise to the phenomenon of weld decay. The metal in this area becomes corroded because it is no longer protected.

## SPOT WELDING

This technique allows thin sheets to be joined by welding from one side only. Thin material can be welded to metal of any thickness quickly and efficiently. Electrodes containing a small amount of thorium to increase their emissivity are often used. Figure 5.19 shows the method of application. The arc gap should be determined by making several test-welds with adjustments to current and time until a satisfactory weld is obtained. Power may be a.c. or d.c. according to the metal to be welded. For thin sheet, manual pressure is used to hold the workpiece close to the arc. For automatic use, an air cylinder will supply mechanical pressure.

The sequence of operations is

(1) gas flow and h.f. initiated by switch (usually foot operated but sometimes situated on the handle of the welding torch)

(2) welding current starts

(3) timing apparatus cuts off current but allows gas to flow a little longer.

Water-cooled torches are necessary for spot welding because of the extra heat generated by the enclosed arc. If the torch is automatically traversed along the surface of the work, a continuous weld can be made; this is a useful technique worthy of further exploitation and development.

## WELDING COPPER

Copper has a high thermal-conductivity, as high as four times that of steel, and massive sections require pre-heating to about 750 °C in order to establish a fluid weldpool. Direct current with electrode negative is most suitable for welding copper. Thoriated tungsten electrodes, because they are more resistant to arc erosion, are recommended; high-purity argon and nitrogen are the normal shielding-gases. Lighter-gauge metal up to about 3 mm thick can be satisfactorily welded using pure argon although a moderate pre-heat may be required. For thicker material, nitrogen is necessary with substantial pre-heating of the workpiece. Nitrogen, being a diatomic gas, dissociates at arc temperature and the heat of recombination adds to the arc heat. Because of the irregularity of the weld surface and the tendency to porosity with nitrogen, argon is sometimes added in the ratio 80 per cent argon, 20 per cent nitrogen.

## WELDING TITANIUM

The metal is comparitively easy to join by the G.S.T.A. technique, the weldpool being fluid and the wetting action good. The most important factor in the welding of titanium and its alloys is satisfactory gas-shielding, since the metal is extremely reactive with atmospheric gases such as oxygen and hydrogen and, to a lesser degree, nitrogen. For manual welding, using pure-argon shielding, the gas shroud must be extended so that the metal is allowed to cool below 500 °C before exposure to air (see figure 5.20). If this is not done, embrittlement of the weld and adjacent areas may occur. Titanium, with 56 per cent of the weight of steel of equal strength, is resistant to many chemical and corrosive environments including sea water, and is increasingly being employed as a substitute for the more conventional stainless-steels.

Direct current with electrode negative is employed for titanium welding and the standard rule of absolute cleanliness of the workpiece, particularly the joint interfaces (which should be cleaned with a titanium-bristle brush) applies. Argon backing of the joint is necessary. Welding in an enclosed cabinet that has been completely exhausted of air then filled with high-purity argon is the best guarantee of executing good welds in titanium. If the workpiece is too large for this, an additional shielding with argon may be necessary. This may be done by building up a wall of asbestos board round the weld area and packing the interstices with fireclay before introducing argon through small-bore pipes, on both sides of the joint. In this improvised method care must be taken not to

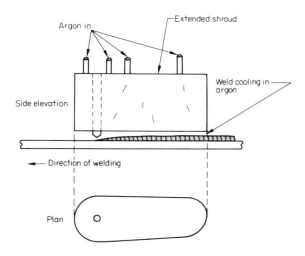

*Figure 5.20    Extended shroud*

induce air by using turbulent-flow conditions of the argon — a condition associated with high pressures and small-bore orifices. Test runs in the environment should therefore be made before welding the actual joint. Titanium pipes can be backed with argon but the extended shroud should be shaped to the pipe diameters.

### Safety with Titanium

Titanium cannot be welded by metal-arc or oxy-acetylene welding and it should be remembered that finely divided dry titanium, including swarf, will burn fiercely if exposed to a naked flame. Water must not be used to extinguish burning titanium — chalk powder or carded asbestos should be used instead.

Welding speed must be kept low to allow the weld metal sufficient time for cooling in the argon. The most satisfactory method is to hold the torch as upright as possible relative to the work surface, and to use autogenous welding of light-gauge material.

The alloys of titanium tend to be more reactive with oxygen than the pure metal and it is imperative that gas shielding is correctly applied. Contaminated zones in titanium and its alloys cannot be repaired unless the whole of the embrittled area has been removed. Titanium is expensive and structures of the metal, because of the special fabricating-techniques employed, are even more so; it is therefore essential that before welding either *in situ* or in an argon-filled cabinet, trial welds are first made on scrap material.

**FURTHER STUDY**

W. V. Binstead and E. G. West, *British Experience in the Argon Arc Welding of Aluminium*, Welding Institute, Cambridge

J. R. Handforth, *Practical Aspects of the Argon Arc Welding of Aluminium Alloys*, Aluminium Development Association, London

B. E. Rossi, *Welding Engineering*, McGraw-Hill, Maidenhead, 1954

B.O.C. Ltd, *Argon Arc Welding*, Waltham Cross, 1960

Brinal Ltd, *Surfacing and Special Electrodes*, Camberley, Surrey (revised annually)

Eutectic Co. Ltd, *Welding Data Book*, Feltham, Middlesex (revised annually)

Sifbronze Ltd, *Specification List* (includes T.I.G., M.I.G. and M.A.G. consumables), Stowmarket, Suffolk (revised annually)

# 6. QUALITY AND INSPECTION

The quality of welded work depends on a number of factors, a few of which are listed below.

1. Good design of structure
2. Correct welding-process selection
3. Suitability of filler material (wire or parent)
4. Precise preparation and alignment of members
5. Suitable jigging that allows the welder good access, visibility and controls distortion
6. The use of a compatible welding-procedure to modify thermal changes
7. Conscientious and skilled workmanship
8. Inspection

The most economical joint for any fusion-welding process is the closed-square butt which can be welded autogenously and is ideal for automatic applications, especially with high-energy processes like electron-beam and laser welding. The upper limit of thickness using this preparation is determined by the power input of the welding process. Oxy-acetylene, for example, is not a satisfactory technique for welding steel plate over 6 mm thick because of the excessive heat input required and the distortion produced. Metal-arc welding, using deep-penetration electrodes extends the range, as does M.A.G. and submerged-arc. To weld really heavy materials, however, processes like electro-slag and electron-beam welding are necessary.

The fillet is also an economical configuration but is more expensive than the closed butt which does not, in many cases, require the addition of filler material. The size and accurate disposition (that is, having equal leg-lengths) of fillet welds is important because: (1) an oversize weld is uneconomical and may cause local stress-concentrations, and (2) an undersize weld may lead to structural failure. Weld dimensions as indicated on the working drawing should therefore be rigidly adhered to, and if pre- or post-heating is indicated, this should be carried out exactly as instructed.

The drawing of the structure should give the following precise information.

1. Process to be used
2. Electrodes, filler wire of flux, etc.
3. Welding positions
4. Welding procedure
5. Size of weld deposit
6. Preparation of joints (machining or flamecutting, etc.)
7. Heat treatment (if any)

Flamecut preparations should be free from fluting or any irregularity and fit-up should be within tolerances. Machine grinding gives the best condition for vee, butt, fillet and 'U' joints but this method is relatively expensive. It is possible to prepare joints by machine flamecutting with a high degree of accuracy. The process is cheaper and faster than machining and is only limited by the type of metal; for example, aluminium, copper and their alloys cannot be prepared in this way — machining or plasma-arc cutting must be used instead. Given satisfactory joint-design, preparation, jigging and fit-up, the result depends ultimately on the welder.

The attributes of a craftsman welder are integrity, skill, ingenuity, good eyesight and a certain amount of intuition. Pride of workmanship is absolutely essential as is the ability to concentrate. Just as handwriting is characteristic of the writer, so does welding mirror craftsmanship — or the lack of it.

## VISUAL INSPECTION

To the experienced eye, many unacceptable weld-defects are easily detected. For example, a minute crater crack, surface porosity (which may denote more extensive sub-surface cavities), lack of fusion (in M.A.G. welding), inequality of leg lengths in fillet welds, irregular deposition, undercut, excessive reinforcement and undersize deposits are some common defects. Destructive testing of welds is expensive and often unnecessary and should only be resorted to if a visual inspection is inconclusive.

Radiographic inspection is also costly and is mostly used

for heavy fabrications designed for arduous service-conditions, for example, pressure vessels or boilers. The interpretation of radiograph calls for a great deal of experience without which there may be some controversy.

Dye-penetrant testing, in which the work is painted with a fluorescent penetrating dye and exposed to ultraviolet light, is another method used.

Magnetic-particle detection in which the workpiece is painted with, or immersed in, a magnetic fluid and then magnetised, is mostly applicable to machined components. The finely divided particles of iron suspended in the liquid are attracted to surface cracks or fissures.

## Simple Aids to Inspection

Simple aids to welding inspection are a pocket magnifier, a steel scriber, a set of feeler gauges and weld-dimension gauges. The magnifier need not be of high power – 3 to 6x is sufficient, and the advantage is that it enables on-the-spot inspection. The scriber is useful for defining and assessing undercut. The weld-dimension gauge can be made from a piece of 3 mm carbon steel or 'silver steel'. A number of such gauges may be necessary to cover numerous weld-sizes adequately. A typical weld-gauge is shown in figure 6.1. The feeler gauges are useful for checking weld-throat concavity with the weld gauge and for measuring surface distortion of the workpiece with a straight edge.

Faults in welding may often be detected with the naked eye but if doubt arises then a macrograph specimen may be used

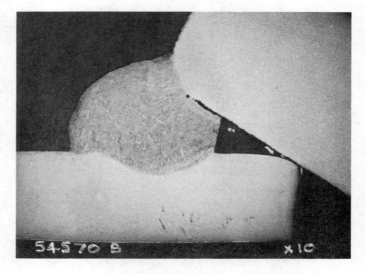

*Figure 6.2    Lack of root fusion ($CO_2$ shielding)*

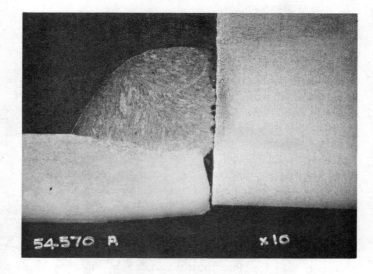

*Figure 6.3    Cold lap ($CO_2$ shielding)*

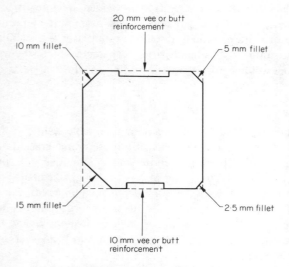

*Figure 6.1    Weld-dimension gauge*

for verification. A macrograph is prepared by cutting, highly polishing and chemically etching a section of the weld and parent metal. This is then magnified by about 10x and photographed. A micrograph is obtained by the same process but much greater magnification is used to show the metallic structure of a selected area in detail. Figure 6.2 shows an M.A.G. weld made with $CO_2$ shielding. The joint shows poor fit-up and lack of root fusion. Figure 6.3 shows a cold lap with

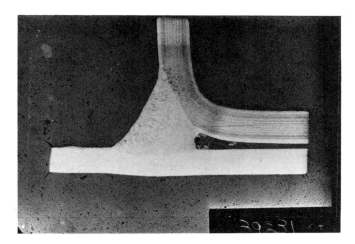

Figure 6.4    Poor joint preparation

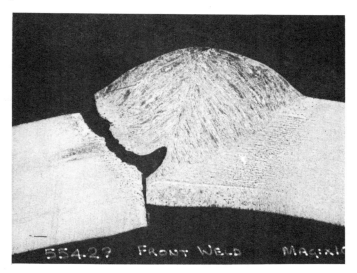

Figure 6.6    Fracture in off-line weld, hardened zone

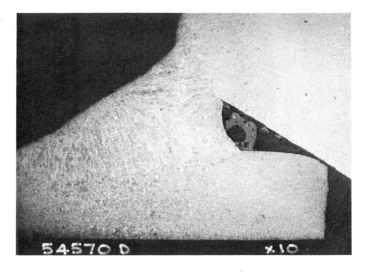

Figure 6.5    Lack of fusion in root ($CO_2$ shielding)

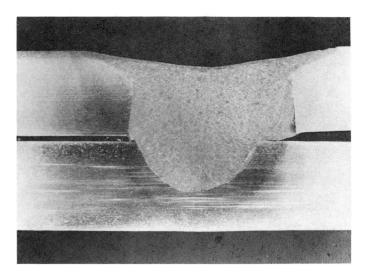

Figure 6.7    Plug weld

even greater lack of fusion on the vertical member together with no root penetration. The weld in figure 6.4 illustrates an M.A.G. fillet in 2 mm steel; this shows good fusion and weld-throat profile. Root fusion with this configuration cannot be effectively assessed. The fillet weld in figure 6.5 shows unequal leg-length, poor joint fit-up and lack of root fusion. An automatic weld is shown in figure 6.6. Here, failure of the weldment has occurred because the weld is grossly off the centre line of the closed butt-joint; lack of penetration is also apparent. A plug weld in 3 mm steel is shown in figure

6.7. The weld shows insufficient penetration of the bottom sheet and some lack of reinforcement on the top sheet. Gross internal porosity is seen in figure 6.8. This was caused by the absence of shielding gas; small micro cracks can also be seen radiating from some of the pores. Good leg-length and side-wall fusion are shown in figure 6.9 but there is a complete lack of root penetration. Weld failure caused by wrong joint-preparation is shown in figure 6.10. This was a weld made in 25 mm thick, low-carbon steel by manual-arc welding.

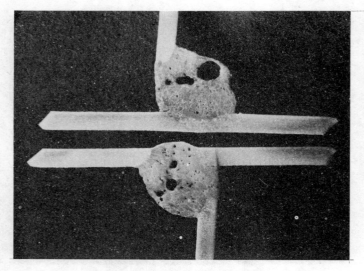

*Figure 6.8    Porosity (no $CO_2$ shielding)*

*Figure 6.10    Weld failure*

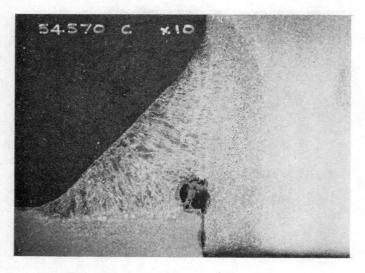

*Figure 6.9    Root porosity ($CO_2$ shielding)*

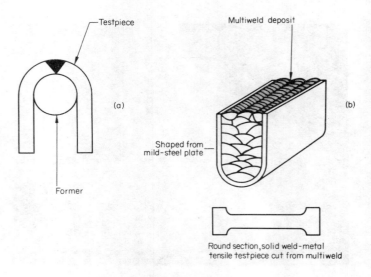

*Figure 6.11    Preparing testpiece*

The preparation required was a single-vee, 70° inclusive angle with 2 mm root gap and 5 mm root face. Instead of this a 'U' groove 10 mm deep and about 15 mm wide was used; the structure subsequently failed early in service.

Although laboratory weld-testing, using sophisticated techniques like X-rays and ultrasonics, is necessary for important welded work, there are simpler methods that can be used in many circumstances. One of these is the bend test (of which there are several variations) – figure 6.11a shows a cross-section of the welded joint bent round a former; this is not an essential requisite, thin sheets can be hammered flat, but this constitutes a very severe test.

To provide a tensile-test specimen (see figure 6.11b) a 'U' preparation is made and then filled by multiple runs. From the deposit the test bar is machined to a specified dimension and stretched in a tensile-testing machine. This method is ideal for the evaluation of metal-arc welding; any voids or inclusions in the waist of the specimen lead to failure below the specified failure-point of the material.

## IMPACT TESTING

Dynamic loading of welded structures demands that welds be capable of sustaining such suddenly applied forces satisfactorily. The Charpy notch impact test is one method of assessing this factor. A similar evaluation is achieved by using the Izod system. In order to be fully proficient, the welding inspector, besides having a thorough knowledge of drawing interpretation, standards required, and the use of checking fixtures and instruments, should preferably have had substantial experience as a practical welder.

## CLASSIFICATION OF DEFECTS

These may be broadly assessed as discontinuities of structure, dimensional faults and variation of physical properties. Discontinuities in weld deposits are such faults as porosity, cracking, insufficient penetration, cold lapping and slag inclusions. Often these faults are detectable by visual inspection and their significance and extent assessed satisfactorily.

Faults of a dimensional nature are usually a result of non-observance of the drawing requirements, misinterpretation of them, lack of communication, poor workmanship, or a combination of these factors.

Variations from specification of physical properties in parent or weld metal is a matter that has to be evaluated by laboratory analysis. Contributory factors can, however, originate from the shop floor. The use of the wrong type of filler material or gross contamination of the workpiece, for example, could produce weld deposits of inferior quality. Test plates are often used to assess the physical properties of parent material and weld metal.

Laboratory testing includes determination of tensile strength, hardness, fatigue resistance, creep, evaluation under service conditions, etc. Large vessels that are required to operate under conditions of high temperature and pressure, are examined by radiography, ultrasonics, or the more conventional pressure-testing.

The soundness of welded joints in small containers such as water tanks or chemical vessels that are not subject to arduous working-requirements can be tested in the following way. The outside of the welded joints are painted with ordinary lime wash (slaked lime and water). After this has thoroughly dried, the insides of the joints are liberally coated with thin paraffin. Any cracks, porosity or discontinuity in the weld or surrounding metal can be seen as regions of the white-washed outer-surface showing paraffin stains. A period of up to eight hours should be allowed to elapse (depending on the wall thickness of the structure) to allow paraffin penetration.

Table 6.1 summarises laboratory testing-methods.

*Table 6.1  Summary of defects revealed by laboratory examination*

| Technique | Type of defect |
|---|---|
| Radiography | Incomplete penetration<br>Inclusions<br>Porosity (spherical or piping)<br>Cracks, but these must be of suitable magnitude and not lie in the plane of the beam<br>Changes of thickness |
| Magnetic | Cracks, especially surface cracks<br>Overlap, discontinuities or voids near to the surface |
| Dye penetrants | Porosity, blowholes (gaseous effusion zones at surface)<br>Piping<br>Cracks |
| Induction, eddy current resistance | Lack of penetration<br>Lack of fusion (cold lapping)<br>Porosity (piping, local or extensive)<br>Sub-surface and surface cracks<br>Inclusions |
| Chemical | Composition defects<br>Specimen preparation (micro and macrographs)<br>Leak detection<br>Corrosion testing |
| Ultrasonic | Incomplete penetration<br>Lack of fusion<br>Porosity<br>Piping, local or extensive<br>Cracks<br>Inclusions<br>Discontinuities |
| Pressure | This method, using air, inert gases, water, other liquids or steam, is used for vessel and pipeline testing to determine strength, work quality and freedom from cracks, porosity or lack or fusion areas |

## STRUCTURAL AND DESIGN ASSESSMENT

Large fabricated structures like pressure vessels, bridges, cranes, ships, locomotives and aircraft are in fact substantial reservoirs of stored energy. This is due mainly to the metal-forming processes and is augmented by welding. Under these circumstances, a crack once started may propagate very rapidly while there is sufficient energy available near the site of origin. Design therefore plays as significant a part in determining the total stress latent in a welded structure as do the welder, welding technique and process. Observance of recommended procedure must be strict and the conscientious craftsman must not tolerate any deviation.

Generally, for structures that must endure cyclic loading the butt joint is superior to the lap joint. Intermittent welding (in the case of an important structure) should be avoided. Welds with any defect or defects in excess of the permitted standard, should be rectified using only specified procedures. Any weld with excessive reinforcement is likely to be an area of high stress-concentration, it is therefore important to keep within the stated dimensions. Extraneous fitments welded to a structure such as brackets, lifting eyes or tapping blocks should be avoided. If this is not possible, careful attention should be given to their siting.

## Stress Relief

The best insurance against dimensional changes during subsequent working (machining or forming) is adequate stress-relieving of a structure.

Welding joints are the focal points of residual, tensile and compressive stresses and these are released in varying degrees, by machining. A good axiom is 'when in doubt, stress relieve'. The application of such heat treatment must be suited to the material, size of structure and service requirements. Time and temperature are the important factors but the provision of sufficient time for the cooling cycle should not be overlooked. Unequal or too rapid cooling may introduce further stresses.

The usual methods of stress relief are: induction heating — a method which gives very precise temperature-control — flame heating and electric- or gas-furnace heating.

## Proof Testing

It is possible for a weld to contain one or more minor defects without being classified as unsatisfactory as long as it does not initiate a source of service failure. The assessment of the weld in such a case can be made by an overload test. The excess loading is applied in the way that it would be under service conditions, but is less than the failure load. In this way proof testing can be used to advantage without destruction or deformation of the structure. Pressure vessels, vehicle chassis, lifting appliances, etc., are some of the items that can be assessed in this manner.

## Jigs

The welding of unrestrained plates will always produce movement from their original location and distortion will result. Tacking in some degree may modify this, and for a one-off job it is not usually economic to construct a jig. The correction of distortion by machining, presswork or heat treatment, however, adds time and cost to the job. Distortion

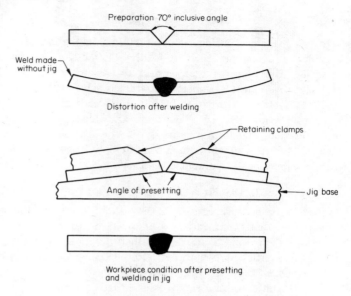

*Figure 6.12    Use of jigs*

can be kept to a minimum and in some cases eliminated by effective fixtures together with a tried and tested welding-procedure. If, for example, it is seen that a weld such as that shown in figure 6.12, produces an upwards inclination of the plate edges, a fixture can be designed to neutralise the effect. To do this, the vee preparation, amount of weld deposited, welding current used, direction of welding and procedure used, should not be varied. It will be seen from trial welds that the amount of distortion occurring in such circumstances is fairly constant. From this can be calculated the amount of presetting in the reverse direction needed to produce a satisfactory weld condition. It does not necessarily follow, however, that because, say, a plate edge has lifted 5 mm from the horizontal after welding, that this amount is required to provide correction. Account has to be taken of other factors such as the inherent properties of the parent material and pre-heat temperature variability. Aluminium, for example, would not require the same amount of pre-setting as high-tensile-steel plate, and cast iron could not be pre-set at all. By using packing pieces or shims in the first instance the ultimate pre-setting angle could be accurately determined. From this information a suitable production-jig could be designed.

An important factor in the designing of jigs and fixtures is the consideration of welder access. It is useless to have a jig that fulfils its purpose but where the joint to be welded cannot be comfortably reached by the welder. Another factor often overlooked is the jig weight. If the welder is required to manipulate a heavy jig several times an hour, his efficiency will deteriorate correspondingly.

The contact faces of the jig where the workpieces are located should be of copper or aluminium bronze. These metals give good electrical-conductivity, act as heat sinks and are spatter resistant if kept cool. One way of avoiding operator fatigue through jig handling is to fix the jig on a rotating table or manipulator; large drums or vessels can be rotated using power-driven rollers as shown in figure 6.13. For automatic welding, special fixtures can be designed incorporating all the necessary process-control requirements on a single panel as shown in figure 6.14. This shows a machine for the automatic M.I.G. welding of small cylindrical components. A more elaborate automatic-welding arrangement is shown in figure 6.15 where multiwelding is being performed using the M.A.G. process.

*Figure 6.13    Large welding-manipulator (courtesy of B.O.C. Murex Ltd)*

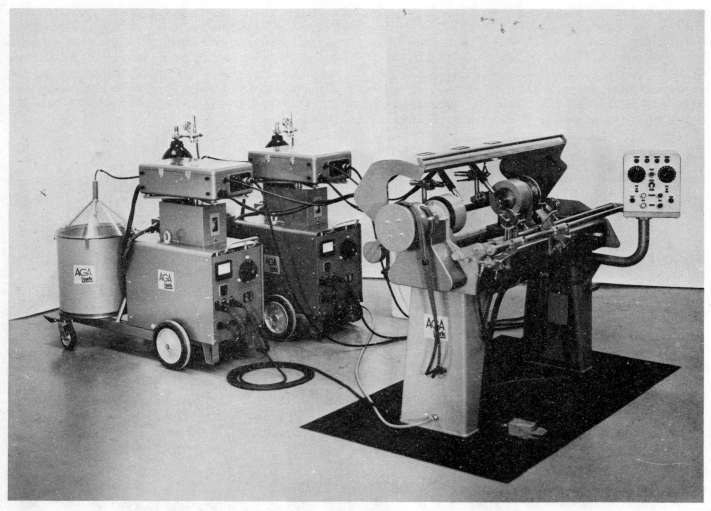

*Figure 6.14    Automatic-welding arrangement (courtesy of A.G.A., U.K., Ltd)*

## SUMMARY OF OTHER WELDING TECHNIQUES AND PROCESSES

### Resistance Welding

This heading covers a large number of associated processes stemming from the original concept of spot welding. Resistance welding is a method of joining two or more pieces of metal by using heat generated by the resistance of the workpieces. Current flowing through low-resistance copper-alloy electrodes and the high-resistance workpiece raises the temperature at the joint interfaces. Increased pressure is then applied to the electrodes by air-operated cylinders, forging the heated areas together to effect a weld — see figure 6.16.

The process is extensively used in sheet-metal fabrications in steel, aluminium and various alloys. The main allied techniques are seam welding (as shown in figure 6.17), stitch, flash, flash butt, projection, roll, spot, mash, stitch and percussion welding.

Because of the extremely accurate time-control required in spot welding, sophisticated equipment has had to be developed. The functions of a control panel are: (1) time control of current flow, (2) magnitude of current, (3) forging time, and (4) synchronisation of air cylinder. The heated areas between

*Figure 6.15    Automatic welding (of legs on to baths) (courtesy of A.G.A., U.K., Ltd)*

the electrode tips must be raised to exactly the right temperature before forging pressure is applied. If the temperature is too high molten metal will be expelled and the quality of the weld will be impaired. The process is ideal for automatic use. In the car industry, for example, multiwelding using large numbers of electrodes operated simultaneously or in sequence is often used. The correct design of electrodes with adequate internal water-cooling is most essential. For welding two sheets of 2 mm steel, for example, a pair of properly designed and cooled electrodes would have an average life of about 5000 spot welds before redressing. Without water cooling it is

doubtful whether they would survive 100 welds. The faces of the electrodes that contact the work must be maintained at the correct diameters. With use, these areas become enlarged and 'mushrooming' occurs. When this happens, current density is affected and weak welds result.

Spot, seam and mash seam welding (figure 6.17) are extensively used in aircraft, motorcars, domestic equipment and many other industries. Water- and pressure-tight joints can quickly be made in a variety of metals and alloys. Power sources for resistance welding are commonly a.c., although d.c. can be used for special applications.

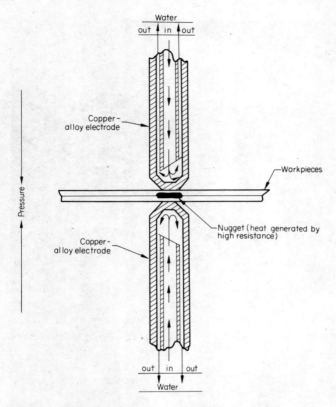

*Figure 6.16    Resistance spot-welding — welding is accomplished by the high resistance of the workpiece to the passage of electric current; forging then takes place by means of pressure applied to the electrodes by air cylinders*

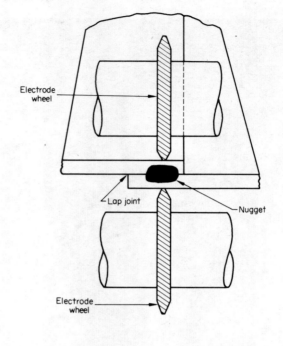

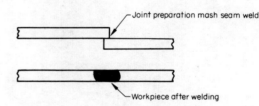

*Figure 6.17    Seam welding*

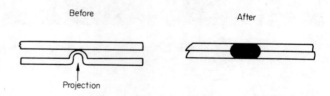

*Figure 6.18    Projection welding*

In flash welding, two surfaces of the workpiece (usually light-gauge metal) connected into the welding circuit are brought first into light contact. A high-magnitude current causes the adjacent workpiece-surfaces to melt and fall away, thus creating a flashing or arcing condition. When the work is sufficiently heated, sudden pressure is applied to effect a welded joint. In flash-butt welding bar sections are joined by bringing the faces into contact and forging them by pressure.

Projection welding is a resistance process in which one member of the workpiece has an accurately dimensioned pip or pressed protuberance on its surface. Figure 6.18 shows one method of using this process. The raised portion of the lower sheet, in this case, is formed by a press operation. Nuts are supplied to industry with projections formed already for welding. The size of projections is most important; if they are too big or too small, the weld quality is severely affected. The sequence of operation is

1. Hold (the electrodes, advanced by the air cylinders, apply light pressure to the work)

2. A low-magnitude current pre-heats the projection which immediately begins to collapse

3. At this point, pressure and welding current are increased.

Platens or slab electrodes are often used in resistance welding, the arrangement consisting of one standard-type top electrode on top and the flat (or shaped to the workpiece contour) electrode underneath. This condition enables 'non-marking' welds to be made, that is, no indentation appears on the platen side of the workpiece.

### Electron-beam Welding

In this process the controlled and focused beam of electrons emanating from a heated tungsten strip (the cathode) is used as a highly concentrated thermal-energy source. The welding head is contained in an evacuated chamber into which the workpiece is also placed. The electron stream is accelerated towards the target, (the joint to be welded) by voltages from 50 to 150 kV. The beam emerging from the source is focused up the work area by using a coil — see figure 6.19. Penetration can be varied by current, voltage acceleration, and

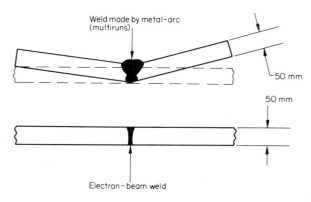

*Figure 6.20    Electron-beam welding reduces distortion*

focusing. Material from 0.1 to 200 mm can be welded with only comparatively local thermal-disturbance of the parent material. Welds made with an electron beam are narrow and have parallel sides, unlike those made with processes previously discussed. Because of the narrow zone of fusion, distortion due to angularity is minimal and no vee-preparation is required, see figure 6.20. Although the electron-beam process is limited by the high vacuum necessary, new techniques of application are increasing the scope of this extremely valuable and efficient technique.

### Friction Welding

As the title implies, this is a form of welding in which the heat necessary to join sections of metal is generated by friction. Cylindrical sections are ideal for joining by this process, one component being rotated against the other in a machine that is a special development of a turning lathe. The heat generated by the friction of the adjacent surfaces first softens the metal in the immediate welding-area under applied pressure, deformation then occurs and plastic metal is squeezed out from the joint, exposing clean uncontaminated surfaces. At this point forging pressure is increased and rotation ceases. The quality of welds produced in this way is extremely high and failure is rare; dissimilar metals can also be welded to each other. A typical friction weld is shown in figure 6.21. The roll-over produced by the expulsion of plastic metal can be seen — this can, if necessary, be machined off. Among the metals and alloys weldable by the process are steel, stainless steel, aluminium, copper and bronze. Because the power requirements are comparitively low and the process is capable of high-volume production it is very economical and is being used on an increasing scale. Some typical applications are motorcar

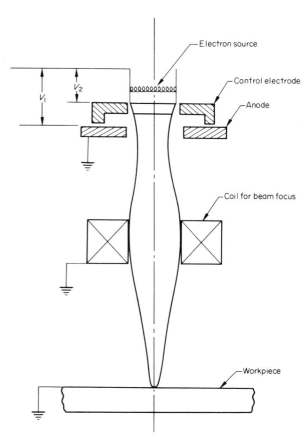

*Figure 6.19    The electron-beam principle*

axles and gears, aircraft components, and pipework. Even fine wires can now be welded as a result of developments by the Institute of Welding.

## Submerged-arc Welding

This process is a method of arc welding using a continuous bare-wire electrode that deposits weld metal beneath a layer of flux. The principle is shown in figure 6.22. The flux, which may be in granular or powder form (determined by the mesh size), can be applied manually or automatically. The process can be used over a wide range of metal thicknesses — from about 2 mm to 130 mm — and in heavy sections one-pass welds can be made in the down-hand or gravity position. The arc, completely hidden by the flux which is fed in advance of the electrode, can be maintained by a voltage feedback system; that is, when the desired operating-parameters have been set on the control panel, the arc length is automatically controlled from the reference (or selected) voltage. A.C. or d.c. power can be used.

The methods of establishing the arc are

1. Electrode feeds forward at a slow speed
2. Short-circuits on the workpiece
3. Electrode direction automatically reversed thus establishing an arc
4. Electrode now feeds downwards again but this time at the speed required for depositing the weld metal
5. Arc-length control is now automatic.

*Figure 6.21    Friction-weld specimen*

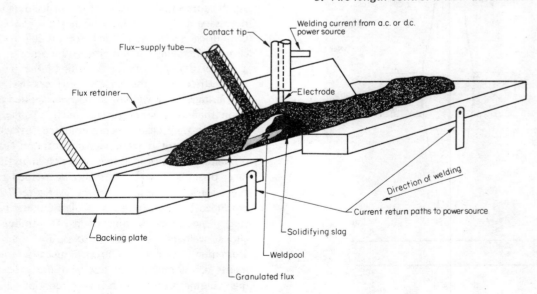

*Figure 6.22    Submerged-arc welding*

Or

1. Electrode feeds downwards to workpiece
2. Short-circuits and burns off creating an arc gap.

Alternatively, the arc may be started by using a high-resistance conductor placed between the electrode tip and workpiece. Steel wool or carbon may be used for this purpose although these methods are rarely used today.

Once the arc is established the growing pool of molten metal melts and displaces the adjacent flux which at arc temperature becomes electrically conductive. Besides its function of protecting the weldpool from atmospheric attack, the flux may contain elements such as manganese, chromium or nickel. In this way the composition of the weld deposit can be matched to that of the parent material, even though the electrode may be low-carbon mild steel. The heavy flux-blanket conserves heat so that submerged-arc welds cool comparitively slowly producing soft ductile deposits. Very large welding-currents and high travel-speeds can be used and penetration and weld profile accurately controlled.

Among the materials that can be welded are low-carbon and medium-carbon steels, heat- and corrosion-resistant steels, and nickel alloys such as stainless steel and Monel. The types of weld possible with submerged arc include butt, plug and fillet welds. A submerged-arc weld in progress is shown in figure 6.23.

Large cylindrical vessels, rotated automatically under a static welding-head are often welded by this process. Thin to thick material, as for example, 2 mm sheet to 10 mm plate, can be accommodated and dissimilar metals can also be welded. Because deep penetration is a marked characteristic of the process, less work-preparation is often required. Butt joints, for example, may be made in steel plate 8 mm thick using a root opening (the space dimension between the vertical faces of the butt joint) of 3 mm with a supporting copper or steel backing-bar. For thicker material a vee preparation not exceeding an inclusive angle of 70° (depending on thickness) can be used with a substantial root face and no root gap. For example, plates 35 mm thick could be prepared as shown in figure 6.24.

Control of penetration with regard to depth and contour is affected by the following.

1. Arc voltage
2. Electrode diameter, electrode travel-speed along the joint
3. Joint preparation
4. Selection of welding-current return-points

*Figure 6.23 Submerged-arc welding (courtesy of E.S.A.B. Ltd)*

5. Electrode angle to workpiece (acute or obtuse angle to the joint longitudinal-line).

A profound difference can be made to weld quality and consistency of deposition by the position and number of current-return leads provided. This is because magnetic flux (magnetic fields) building up in the jig and machine can cause deflection of the arc. Frequent resiting of the welding-current return-cable is often required on mass-production machines because they tend to become magnetically polarised by repeating the same sequence of operations. Often two or more return leads are used to combat this condition (known also as 'arc blow'). Because the arc is invisible, the detection of this condition is not easy unless the welder is aware of the symptoms. The main ones are

1. Weld deposit deviation from the joint line
2. Variable penetration

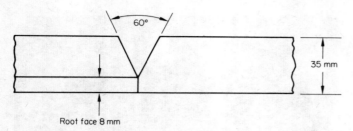

*Figure 6.24    Single-vee preparation*

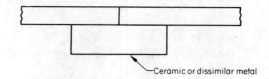

*Figure 6.25    Non-fusible backing*

3. Local areas of porosity that are repeated with each weld run in the same location on the joint

4. Unsteady needle on the voltmeter.

There are also other reasons for producing the symptoms mentioned but any assessment of malfunctioning requires knowledge of other relevant factors. For example, work contamination may cause porosity. Slipping or incorrectly adjusted feed-rolls may be reflected in erratic deposition and unsteady voltmeter-readings. Variable penetration can be produced by variations in root gap, etc.

Another adaptation of the process is twin-electrode welding or series welding in which two electrodes are simultaneously fed into the weldpool. One electrode is connected to the positive pole of the generator output and the other to the negative pole. This system enables low-penetration welds to be made with very little dilution of the parent metal. It finds most use in cladding with dissimilar metals. A.C. or d.c. power can be used with series welding, a.c. generally for steels and d.c. for non-ferrous materials.

Backing methods for submerged-arc welding are

1. Non-fusible backing as shown in figure 6.25
2. Fusible backing that becomes a part of the weldment, that is, a backing strip of the parent material, steel, stainless steel, etc.
3. Non-fusible backing such as massive copper sections
4. Flux backing, in which flux contained in a trough beneath the joint to be welded is rammed against the workpiece by inflating a flexible air-pipe running directly along the joint line.

Flux from submerged-arc welding can be cleaned of contaminants, regraded for particle size and fed back into the welding machine. Some fluxes are hygroscopic and should therefore always be stored and used in dry conditions because water can cause gross porosity in submerged-arc welds. The composition of the weld metal may be varied by adding ingredients to the flux. Metals like chromium, manganese, and

nickel, for example, may be included to maintain or modify the physical properties of the deposit. Therefore using a low-carbon-steel electrode wire, a wide range of deposits are possible, each with its required physical and metallurgical properties.

### Plasma-gas Metal-arc Welding

A new process combining a continuously fed electrode and plasma arc has been developed. The tungsten electrode is not concentric with the gas shroud so that a contact tip can also be accommodated axially. Argon is used as a plasma gas, the external shielding-gas being argon, carbon dioxide or mixtures. The consumable electrode is fed through the plasma gas before entering the secondary gas-zone.

The non-consumable tungsten electrode and the contact tip are water-cooled and the transferred plasma-arc is used, with electrode positive. An interesting feature of the process is the rotating arc that revolves at about 200 r.p.m. This means that thermal energy is directed over a greater area than with the usually stiff columnar plasma-jet and so flat beads with low penetration are obtained. Conversely, the conventional plasma-arc conditions can be achieved by reducing the current passing through the consumable-wire electrode. If this is done, the plasma jet is restricted, giving deep narrow penetration.

### Plasma-cutting (Transferred Arc)

Using a high-powered transferred arc ferrous and non-ferrous metals can be cut at fast speeds. Copper and stainless steel, for example, can be cut in thicknesses exceeding 100 mm. In ship-building, where large steel plates have to be cut, plasma-cutting with computer control is an attractive proposition. For miscellaneous small items, however, the computer-directed multihead oxy-acetylene profiler is more economical.

### Plasma-plating

By introducing high melting-point powdered metals like tungsten into the plasma arc, components can be produced that were formerly impossible to make. The object is first made out of steel or other suitable material. It is then plasma-sprayed until the desired thickness has been obtained, and the mould or pattern is dissolved by chemical action.

## Plasma-cutting (Non-transferred Arc)

Non-electrically conductive materials, like nylon, can be cut using this mode. Since a greater proportion of the thermal energy is confined between the two electrodes in the torch, this system is somewhat restricted in application.

## Explosive Welding

Controlled explosions are used in this process to bring plates (often of dissimilar metals) into contact under pressure. This technique is an extension of the cladding process and is still being developed.

## Vapour Plating

Dissimilar metals can be joined by coating with gold, tantalum and other metals and then heating the parts in a reducing atmosphere. Diffusion of the interface metals occurs by lattice and grain-boundary modifications (p. 8).

## Vacuum Welding

The absence of a gaseous environment offers a great advantage in metal joining and even non-metals like ceramics and graphite can be bonded in this way. In addition, high-purity metal can be produced by vacuum refining. (Electron-beam welding is a vacuum process.)

The frontiers of metal joining are continually expanding with the gathering momentum of technological advancement. Such processes as the laser beam (for welding and cutting), explosion welding, diffusion bonding, plasma-arc welding and cutting, along with many more, are not within the scope of this book. Indeed, at this point in the study of the basic principles of fusion, such diversions may only confuse the reader. References can, however, be made to various information sources for the purpose of further study — notably the Welding Institute, which provides (among many other services) testing of welds and materials, advice on equipment selection and expert help with the design of welded fabrications.

# FURTHER STUDY

## General

BS 499:Part 1:1965 Welding, brazing and thermal cutting glossary
Part 2:1965 Symbols for welding
BS 2929:1957 Safety colours for use in industry

## Processes

BS 693:1960 General requirements for oxyacetylene welding of mild steel
BS 1140:1957 General requirements for spot welding of light assemblies in mild steel
BS 1723:1963 Brazing
BS 1724:1959 Bronze welding by gas
BS 1821:1957 Class I arc welding of steel pipelines and pipe assemblies for carrying fluids
BS 2633:1973 Class I arc welding of ferritic steel pipework for carrying fluids
BS 2971:1961 Class II metal-arc welding of steel pipelines and pipe assemblies for carrying fluids
BS 3019:General recommendations for manual inert-gas tungsten-arc welding
Part 1:1958 Wrought aluminium, aluminium alloys and magnesium alloys;
Part 2:1960 Austenitic stainless and heat-resisting steels
BS 3571:General recommendations for manual inert-gas metal-arc welding
Part 1:1962 Aluminium and aluminium alloys
BS 4570:Fusion welding of steel castings; Part 1:1970 Production, rectification and repair; Part 2:1972 Fabrication welding
BS 4677:1971 Class I arc welding of austenitic stainless steel pipework for carrying fluids

## Equipment

BS 638:1966 Arc welding plant, equipment and accessories

## Welding Consumables

BS 639:1972 Covered electrodes for the manual metal-arc welding of mild steel and medium tensile steel
BS 1453:1972 Filler metals for gas welding
BS 1719:Classification, coding and marking of covered electrodes for metal-arc welding; Part 1:1969 Classification and coding

BS 1845:1966 Filler metals for brazing

BS 2901:Filler rods and wires for gas-shielded arc welding; Part 1:1970 Ferritic steels; Part 2:1970 Austenitic stainless steels; Part 3:1970 Copper and copper alloys; Part 4:1970 Aluminium and aluminium alloys and magnesium alloys; Part 5:1970 Nickel and nickel alloys

BS 2926:1970 Chromium-nickel austenitic and chromium steel electrodes for manual metal-arc welding

BS 4165:1971 Electrode wires and fluxes for the submerged-arc welding of carbon steel and medium-tensile steel

## Testing and Inspection

BS 709:1971 Methods of testing fusion welded joints and weld metal in steel — metric units

BS 1295:1959 Tests for use in the training of welders. Manual metal-arc and oxy-acetylene welding of mild steel

BS 2600:Methods for the radiographic examination of fusion welded butt joints in steel; Part 1:1973 5 mm up to and including 50 mm thick; Part 2:1973 Over 50 mm up to and including 200 mm thick

BS 2910:1973 Methods for the radiographic examination of fusion welded circumferential butt joints in steel pipes

BS 3923:Methods for ultrasonic examination of welds; Part 1:1968 Manual examination of fusion butt joints in ferritic steels; Part 2:1972 Automatic examination of fusion welded butt joints in ferritic steels; Part 3:1972 Manual examination of nozzle welds

BS 4069:1966 Magnetic flaw detection inks and powders

BS 4206:1967 Methods of testing fusion welds in copper and copper alloys

BS 4396:1969 Methods for magnetic particle testing of welds

BS 4416:1969 Method for penetrant testing of welded or brazed joints in metals

BS 4870:Approval testing of welding procedures; Part 1:1974 Fusion welding of steel

BS 4871:Approval testing or welders working to approved welding procedures; Part 1:1974 Fusion welding of steel

BS 4872:Approval testing of welders when welding procedure approval is not required; Part 1:1972 Fusion welding of steel

# Appendix

## UNITS AND SYMBOLS

Symbols for quantities are in *italic* type, and abbreviations for the names of units are in ordinary type

| | | | | |
|---|---|---|---|---|
| A | ampere | | kl | kilolitre |
| A/cm$^2$ | ampere per square centimetre | | kV | kilovolt |
| A/in.$^2$ | ampere per square inch | | kVA | kilovolt ampere |
| atm | atmosphere | | kW | kilowatt |
| Btu | British thermal unit | | kWh | kilowatt hour |
| cal | calorie | | lb | pound |
| cm | centimetre | | lb/in.$^2$ | pound per square inch |
| cm$^2$ | square centimetre | | MHz | megahertz ($10^6$ Hz) |
| cm/min | centimetre per minute | | m | metre |
| fl. oz | fluid ounce | | min | minute |
| ft | foot | | ml | millilitre |
| ft$^3$ | cubic foot | | mm | millimetre |
| g | gram | | mm Hg | millimetre of mercury |
| GHz | gigahertz ($10^9$ Hz) | | ms | millisecond |
| gal | gallon | | m/s$^2$ | metre per second per second |
| gr | grain | | N | newton |
| Hz | hertz (cycle per second) | | *P* | power |
| h | hour | | *R* | resistance |
| *I* | electric current | | s, sec | second |
| in. | inch | | *T* | time |
| in.$^2$ | square inch | | V | volt |
| in./min | inch per minute | | *V* | voltage |
| J | joule | | VA | volt ampere |
| kcal | kilocalorie | | W | watt |
| kg | kilogram | | Ω | ohm |
| kg/cm$^2$ | kilogram per square centimetre | | μF | microfarad ($10^{-6}$ F) |
| kHz | kilohertz ($10^3$ Hz) | | μΩ | microhm ($10^{-6}$ Ω) |
| kJ | kilojoule | | | |

## CONVERSION FACTORS

|  | Multiply by | Reciprocal |
|---|---|---|
| Inches to millimetres | 25.4 | 0.039 |
| Feet to metres | 0.305 | 3.281 |
| Yards to metres | 0.914 | 1.094 |
| Miles to kilometres | 1.609 | 0.621 |
| Sq. inches to sq. millimetres | 645.16 | 0.0016 |
| Sq. inches to sq. centimetres | 6.452 | 0.155 |
| Sq. feet to sq. metres | 0.093 | 10.764 |
| Cu. inches to cu. centimetres | 16.387 | 0.061 |
| Cu. feet to cu. metres | 0.028 | 35.315 |
| Pounds to kilograms | 0.453 | 2.205 |
| Tons to tonnes | 1.016 | 0.984 |
| Foot-pounds to kilogram metres | 0.138 | 7.233 |
| Pounds per sq. inch to kilograms per sq. centimetre | 0.070 | 14.223 |
| Pounds per sq. inch to atmospheres | 0.068 | 14.696 |
| Pints to litres | 0.568 | 1.76 |
| Gallons to litres | 4.546 | 0.22 |
| Watts to horsepower | 0.00134 | 745.7 |
| Watts to metric horsepower | 0.00136 | 735.5 |

### Temperatures

Fahrenheit to Centigrade: subtract 32, then multiply by 5/9.
Centigrade to Fahrenheit: multiply by 9/5, then add 32.

### Heat Units

The *British thermal unit* (Btu) is the heat required to raise the temperature of 1 lb of water by 1°F. The *therm* is equal to 100 000 Btu.

The *kilocalorie* (kcal) is the heat required to raise the temperature of 1 kg of water by 1 °C.

The basic unit of energy, whether it be mechanical energy (work), electrical energy or heat, is the *joule* (J). A more convenient unit is often the *kilojoule* (1 kJ = 1000 J).

     1 Btu = 1.055 kJ
     1 kcal = 4.187 kJ

A joule is a *watt second*. A *kilowatt hour* (kWh) (commonly known as a 'unit') is $10^3$ x 60 x 60 J, or 3600 kJ, so that

     1 kWh = 3412 Btu = 860 kcal

## CONVERSION TABLES

1 inch = 25.399978 millimetres; 1 metre = 39.370113 inches.

### Inches to Millimetres

*Fractions*

| | | in. | | mm |
|---|---|---|---|---|
| | | 1/64 | .015625 | .3969 |
| | 1/32 | | .031250 | .7937 |
| | | 3/64 | .046875 | 1.1906 |
| 1/16 | | | .062500 | 1.5875 |
| | | 5/64 | .078125 | 1.9844 |
| | 3/32 | | .093750 | 2.3812 |
| | | 7/64 | .109375 | 2.7781 |
| 1/8 | | | .125000 | 3.1750 |
| | | 9/64 | .140625 | 3.5719 |
| | 5/32 | | .156250 | 3.9687 |
| | | 11/64 | .171875 | 4.3656 |
| 3/16 | | | .187500 | 4.7625 |
| | | 13/64 | .203125 | 5.1594 |
| | 7/32 | | .218750 | 5.5562 |
| | | 15/64 | .234375 | 5.9531 |
| 1/4 | | | .250000 | 6.3500 |
| | | 17/64 | .265625 | 6.7469 |
| | 9/32 | | .281250 | 7.1437 |
| | | 19/64 | .296875 | 7.5406 |
| 5/16 | | | .312500 | 7.9375 |
| | | 21/64 | .328125 | 8.3344 |
| | 11/32 | | .343750 | 8.7312 |
| | | 23/64 | .359375 | 9.1281 |
| 3/8 | | | .375000 | 9.5250 |
| | | 25/64 | .390625 | 9.9219 |
| | 13/32 | | .406250 | 10.3187 |
| | | 27/64 | .421875 | 10.7156 |
| 7/16 | | | .437500 | 11.1125 |
| | | 29/64 | .453125 | 11.5094 |
| | 15/32 | | .468750 | 11.9062 |
| | | 31/64 | .484375 | 12.3031 |
| 1/2 | | | .500000 | 12.7000 |
| | | 33/64 | .515625 | 13.0969 |
| | 17/32 | | .531250 | 13.4937 |
| | | 35/64 | .546875 | 13.8906 |
| 9/16 | | | .562500 | 14.2875 |
| | | 37/64 | .578125 | 14.6844 |
| | 19/32 | | .593750 | 15.0812 |
| | | 39/64 | .609375 | 15.4781 |
| 5/8 | | | .625000 | 15.8750 |
| | | 41/64 | .640625 | 16.2719 |
| | 21/32 | | .656250 | 16.6687 |
| | | 43/64 | .671875 | 17.0656 |
| 11/16 | | | .687500 | 17.4625 |
| | | 45/64 | .703125 | 17.8594 |
| | 23/32 | | .718750 | 18.2562 |
| | | 47/64 | .734375 | 18.6531 |
| 3/4 | | | .750000 | 19.0500 |
| | | 49/64 | .765625 | 19.4469 |
| | 25/32 | | .781250 | 19.8437 |
| | | 51/64 | .796875 | 20.2406 |
| 13/16 | | | .812500 | 20.6375 |
| | | 53/64 | .828125 | 21.0344 |
| | 27/32 | | .843750 | 21.4312 |

## Fractions (continued)

| | | | in. | mm |
|---|---|---|---|---|
| 7/8 | | 55/64 | .859375 | 21.8281 |
| | | | .875000 | 22.2250 |
| | | 57/64 | .890625 | 22.6219 |
| | 29/32 | | .906250 | 23.0187 |
| | | 59/64 | .921875 | 23.4156 |
| 15/16 | | | .937500 | 23.8125 |
| | | 61/64 | .953125 | 24.2094 |
| | 31/32 | | .968750 | 24.6062 |
| | | 63/64 | .984375 | 25.0031 |

| 1/1000 in. | | 1/100 in. | | 1/10 in. | |
|---|---|---|---|---|---|
| in. | mm | in. | mm | in. | mm |
| .001 | .0254 | .01 | .254 | .1 | 2.54 |
| .002 | .0508 | .02 | .508 | .2 | 5.08 |
| .003 | .0762 | .03 | .762 | .3 | 7.62 |
| .004 | .1016 | .04 | 1.016 | .4 | 10.16 |
| .005 | .1270 | .05 | 1.270 | .5 | 12.70 |
| .006 | .1524 | .06 | 1.524 | .6 | 15.24 |
| .007 | .1778 | .07 | 1.778 | .7 | 17.78 |
| .008 | .2032 | .08 | 2.032 | .8 | 20.32 |
| .009 | .2286 | .09 | 2.286 | .9 | 22.86 |

### Units

| in. | mm | 10 | 20 | 30 |
|---|---|---|---|---|
| 0 | | 254.0 | 508.0 | 762.0 |
| 1 | 25.4 | 279.4 | 533.4 | 787.4 |
| 2 | 50.8 | 304.8 | 558.8 | 812.8 |
| 3 | 76.2 | 330.2 | 584.2 | 838.2 |
| 4 | 101.6 | 355.6 | 609.6 | 863.6 |
| 5 | 127.0 | 381.0 | 635.0 | 889.0 |
| 6 | 152.4 | 406.4 | 660.4 | 914.4 |
| 7 | 177.8 | 431.8 | 685.8 | 939.8 |
| 8 | 203.2 | 457.2 | 711.2 | 965.2 |
| 9 | 228.6 | 482.6 | 736.6 | 990.6 |

## Millimetres to Inches

### Units

| | mm | 10 | 20 | 30 | 40 | 50 |
|---|---|---|---|---|---|---|
| 0 | | .39370 | .78740 | 1.18110 | 1.57840 | 1.96851 |
| 1 | .03937 | .43307 | .82677 | 1.22047 | 1.61417 | 2.00788 |
| 2 | .07874 | .47244 | .86615 | 1.25984 | 1.65354 | 2.04725 |
| 3 | .11811 | .51181 | .90551 | 1.29921 | 1.69291 | 2.08662 |
| 4 | .15748 | .55118 | .94488 | 1.33858 | 1.73228 | 2.12599 |
| 5 | .19685 | .59055 | .98425 | 1.37795 | 1.77165 | 2.16536 |
| 6 | .23622 | .62992 | 1.02362 | 1.41732 | 1.81103 | 2.20473 |
| 7 | .27559 | .66929 | 1.06299 | 1.45669 | 1.85040 | 2.24410 |
| 8 | .31496 | .70866 | 1.10236 | 1.49606 | 1.88977 | 2.28347 |
| 9 | .35433 | .74803 | 1.14173 | 1.53543 | 1.92914 | 2.32284 |

| | 60 | 70 | 80 | 90 |
|---|---|---|---|---|
| 0 | 2.36221 | 2.75591 | 3.14961 | 3.54331 |
| 1 | 2.40158 | 2.79528 | 3.18898 | 3.58268 |
| 2 | 2.44095 | 2.83465 | 3.22835 | 3.62205 |
| 3 | 2.48032 | 2.87402 | 3.26772 | 3.66142 |
| 4 | 2.51969 | 2.91339 | 3.30709 | 3.70079 |
| 5 | 2.55906 | 2.95276 | 3.34646 | 3.74016 |
| 6 | 2.59843 | 2.99213 | 3.38583 | 3.77953 |
| 7 | 2.63780 | 3.03150 | 3.42520 | 3.81890 |
| 8 | 2.67717 | 3.07087 | 3.46457 | 3.85827 |
| 9 | 2.71654 | 3.11024 | 3.50934 | 3.89764 |

### Fractions

| 1/1000 mm | | 1/100 mm | | 1/10 mm | |
|---|---|---|---|---|---|
| mm | in. | mm | in. | mm | in. |
| 0.001 | .000039 | 0.01 | .00039 | 0.1 | .00394 |
| 0.002 | .000079 | 0.02 | .00079 | 0.2 | .00787 |
| 0.003 | .000118 | 0.03 | .00118 | 0.3 | .01181 |
| 0.004 | .000157 | 0.04 | .00157 | 0.3 | .01575 |
| 0.005 | .000197 | 0.05 | .00197 | 0.5 | .01969 |
| 0.006 | .000236 | 0.06 | .00236 | 0.6 | .02362 |
| 0.007 | .000276 | 0.07 | .00276 | 0.7 | .02756 |
| 0.008 | .000315 | 0.08 | .00315 | 0.8 | .03150 |
| 0.009 | .000354 | 0.09 | .00354 | 0.9 | .03543 |

## Units (continued)

|     | mm | 100 | 200 | 300 |
|-----|------|---------|---------|---------|
| 0   |        | 3.93701 | 7.87402 | 11.8110 |
| 10  | .39370 | 4.33071 | 8.26772 | 12.2047 |
| 20  | .78740 | 4.72411 | 8.66142 | 12.5984 |
| 30  | 1.18110 | 5.11811 | 9.05513 | 12.9921 |
| 40  | 1.57840 | 5.51181 | 9.44883 | 13.3858 |
| 50  | 1.96851 | 5.90552 | 9.84252 | 13.7795 |
| 60  | 2.36221 | 6.29922 | 10.23620 | 14.1732 |
| 70  | 2.75591 | 6.69292 | 10.62990 | 14.5669 |
| 80  | 3.14961 | 7.08662 | 11.02360 | 14.9606 |
| 90  | 3.54331 | 7.48032 | 11.41730 | 15.3543 |

|     | 400 | 500 | 600 |
|-----|---------|---------|---------|
| 0   | 15.7480 | 19.6851 | 23.6221 |
| 10  | 16.1417 | 20.0788 | 24.0158 |
| 20  | 16.5354 | 20.4725 | 24.4095 |
| 30  | 16.9291 | 20.8662 | 24.8032 |
| 40  | 17.3228 | 21.2599 | 25.1969 |
| 50  | 17.7165 | 21.6536 | 25.5906 |
| 60  | 18.1103 | 22.0473 | 25.9843 |
| 70  | 18.5040 | 22.4410 | 26.3780 |
| 80  | 18.8977 | 22.8347 | 26.7717 |
| 90  | 19.2914 | 23.2284 | 27.1654 |

|     | 700 | 800 | 900 |
|-----|---------|---------|---------|
| 0   | 27.5591 | 31.4961 | 35.4331 |
| 10  | 27.9528 | 31.8898 | 35.8268 |
| 20  | 28.3465 | 32.2835 | 36.2205 |
| 30  | 28.7402 | 32.6772 | 36.6142 |
| 40  | 29.1339 | 33.0709 | 37.0079 |
| 50  | 29.5276 | 33.4646 | 37.4016 |
| 60  | 29.9213 | 33.8583 | 37.7953 |
| 70  | 30.3150 | 34.2520 | 38.1890 |
| 80  | 30.7087 | 34.6457 | 38.5827 |
| 90  | 31.1024 | 35.0394 | 38.9764 |

## COMPARATIVE DIAMETERS

| Diameter | in. | S.W.G. | mm |
|----------|--------|--------|------|
| 1/2 | 0.500 | 7/0 | 12.7 |
| 15/32 | 0.464 | 6/0 | 11.8 |
| 7/16 | 0.432 | 5/0 | 11.0 |
| 13/32 | 0.400 | 4/0 | 10.2 |
| 3/8 | 0.372 | 3/0 | 9.4 |
| 11/32 | 0.348 | 2/0 | 8.8 |
|  | 0.324 | 1/0 | 8.2 |
|  | 0.300 | 1 | 7.6 |
|  | 0.276 | 2 | 7.0 |
| 1/4 | 0.252 | 3 | 6.4 |
|  | 0.232 | 4 | 5.9 |
|  | 0.212 | 5 | 5.4 |
| 3/16 | 0.192 | 6 | 4.9 |
|  | 0.176 | 7 | 4.5 |
| 5/32 | 0.160 | 8 | 4.1 |
|  | 0.144 | 9 | 3.7 |
| 1/8 | 0.128 | 10 | 3.3 |
|  | 0.116 | 11 | 3.0 |
|  | 0.104 | 12 | 2.6 |
| 3/32 | 0.092 | 13 | 2.3 |
|  | 0.080 | 14 | 2.0 |
|  | 0.072 | 15 | 1.8 |
| 1/16 | 0.064 | 16 | 1.6 |
|  | 0.056 | 17 | 1.4 |
| 3/64 | 0.048 | 18 | 1.2 |
|  | 0.040 | 19 | 1.0 |
|  | 0.036 | 20 | 0.9 |
| 1/32 | 0.032 | 21 | 0.8 |
|  | 0.028 | 22 | 0.7 |
|  | 0.024 | 23 | 0.6 |
|  | 0.022 | 24 | 0.55 |
|  | 0.020 | 25 | 0.5 |
|  | 0.018 | 26 | 0.45 |
|  | 0.0164 | 27 | 0.4 |
| 1/64 | 0.0148 | 28 | 0.37 |
|  | 0.0136 | 29 | 0.35 |
|  | 0.0124 | 30 | 0.30 |

## TRAVEL-SPEED CONVERSION TABLE

### Welding speeds

| m/h | ft/h | in./min | m/h | ft/h | in./min |
|-----|------|---------|-----|------|---------|
| 10 | 33 | 7 | 120 | 394 | 79 |
| 20 | 66 | 13 | 140 | 459 | 92 |
| 30 | 98 | 20 | 160 | 525 | 105 |
| 40 | 131 | 26 | 180 | 590 | 118 |
| 50 | 164 | 33 | 200 | 656 | 131 |
| 60 | 197 | 39 | 220 | 722 | 144 |
| 70 | 230 | 46 | 240 | 787 | 157 |
| 80 | 262 | 52 | 260 | 853 | 170 |
| 90 | 295 | 59 | 280 | 918 | 183 |
| 100 | 328 | 65 | 300 | 984 | 197 |

## TENSILE-STRESS EQUIVALENTS

| ton/in.$^2$ | kg/mm$^2$ | p.s.i. | ton/in.$^2$ | kg/mm$^2$ | p.s.i. |
|--------|--------|--------|--------|--------|--------|
| 10 | 15.75 | 22 400 | 26 | 40.95 | 58 240 |
| 11 | 17.33 | 24 640 | 27 | 42.53 | 60 480 |
| 12 | 18.90 | 26 880 | 28 | 44.10 | 62 720 |
| 13 | 20.48 | 29 120 | 29 | 45.68 | 64 960 |
| 14 | 22.05 | 31 360 | 30 | 47.25 | 67 200 |
| 15 | 23.63 | 33 600 | 31 | 48.83 | 69 440 |
| 16 | 25.20 | 35 840 | 32 | 50.40 | 71 680 |
| 17 | 26.78 | 38 080 | 33 | 51.98 | 73 920 |
| 18 | 28.35 | 40 320 | 34 | 53.55 | 76 160 |
| 19 | 29.93 | 42 560 | 35 | 55.14 | 78 400 |
| 20 | 31.50 | 44 800 | 36 | 56.70 | 80 640 |
| 21 | 33.08 | 47 040 | 37 | 58.28 | 82 880 |
| 22 | 34.66 | 49 280 | 38 | 59.85 | 85 120 |
| 23 | 36.23 | 51 250 | 39 | 61.43 | 87 380 |
| 24 | 37.80 | 53 760 | 40 | 63.00 | 89 600 |
| 25 | 39.38 | 56 000 |  |  |  |

## APPROXIMATE FLAME-TEMPERATURES

|  | °C |  | °C |
|---|---|---|---|
| oxy-acetylene | 3100–3300 | air—acetylene | 2460 |
| oxy-propane | 2500 | air—coal-gas | 1871 |
| oxy-hydrogen | 2370 | air—propane | 1750 |
| oxy-coal-gas | 2200 |  |  |

## SUMMARY OF PROCESS POWER-SOURCES

|  | G.S.T.A. | Manual metal-arc | Submerged-arc | G.S.M.A. |
|---|---|---|---|---|
| Power-source characteristics | Drooping | Drooping | Drooping (or flat) | Flat |
| Arc length controlled by | Operator | Operator | Arc voltage via servo-mechanism (or voltage of power source) | Voltage of power source |
| Arc current set by | Power source | Power source | Power source (or wire-feed speed) | Wire-feed speed |

## PROPERTIES OF METALS

| Element | | Melting point (°C) | Specific heat relative to water | Specific gravity (g/ml) | Coefficient of linear expansion (centigrade) (x 10⁻⁶) | Atomic weight |
|---|---|---|---|---|---|---|
| Aluminium | Al | 660.1 | 0.219 | 2.7 | 23 | 26.98 |
| Antimony | Sb | 630.5 | 0.049 | 6.7 | 11 | 121.76 |
| Barium | Ba | 710 | 0.067 | 3.5 | very small | 137.36 |
| Beryllium | Be | 1280 | 0.425 | 1.8 | 12.1 | 9.01 |
| Bismuth | Bi | 271.3 | 0.031 | 9.8 | 13 | 209.0 |
| Cadmium | Cd | 320.9 | 0.056 | 8.65 | 30 | 112.41 |
| Caesium | Cs | 28.6 | 0.052 | 1.9 | 94 | 132.91 |
| Calcium | Ca | 850 | 0.158 | 1.55 | 25 | 40.08 |
| Cerium | Ce | 804 | 0.051 | 6.9 | very small | 140.13 |
| Chromium | Cr | 1900 | 0.12 | 7.1 | 7 | 52.01 |
| Cobalt | Co | 1492 | 0.102 | 8.9 | 12 | 58.94 |
| Copper | Cu | 1083 | 0.093 | 8.96 | 16.7 | 63.54 |
| Gold | Au | 1063 | 0.031 | 19.3 | 14 | 197.0 |
| Iridium | Ir | 2443 | 0.031 | 22.4 | 6.5 | 192.2 |
| Iron | Fe | 1539 | 0.108 | 7.9 | 11.7 | 55.85 |
| Lead | Pb | 327.3 | 0.031 | 11.3 | 29 | 207.21 |
| Magnesium | Mg | 650 | 0.245 | 1.7 | 25 | 24.32 |
| Manganese | Mn | 1250 | 0.112 | 7.4 | 12.8 | 54.94 |
| Mercury | Hg | −38.87 | 0.033 | 13.55 | 182 | 200.61 |
| Molybdenum | Mo | 2620 | 0.065 | 10.2 | 5 | 95.95 |
| Nickel | Ni | 1453 | 0.108 | 8.9 | 12.8 | 58.71 |
| Platinum | Pt | 1769 | 0.032 | 21.45 | 8.9 | 195.09 |
| Potassium | K | 63.2 | 0.177 | 0.86 | 83 | 39.1 |
| Rhodium | Rh | 1960 | 0.059 | 12.4 | 8.4 | 102.91 |
| Silver | Ag | 960.8 | 0.056 | 10.5 | 19 | 107.88 |
| Sodium | Na | 97.8 | 0.296 | 0.97 | 72 | 22.99 |
| Strontium | Sr | 770 | 0.074 | 2.6 | very small | 87.63 |
| Tantalum | Ta | 3000 | 0.036 | 16.6 | 6 | 180.95 |
| Tellurium | Te | 450 | 0.048 | 6.2 | 17 | 127.61 |
| Tin | Sn | 231.9 | 0.056 | 7.3 | 21 | 118.70 |
| Titanium | Ti | 1680 | 0.126 | 4.5 | 9 | 47.9 |
| Tungsten | W | 3380 | 0.034 | 19.3 | 4.5 | 183.86 |
| Uranium | U | 1133 | 0.028 | 19.05 | very small | 238.07 |
| Vanadium | V | 1920 | 0.115 | 6.1 | very small | 50.95 |
| Zinc | Zn | 419.5 | 0.094 | 7.1 | 30 | 65.38 |
| Zirconium | Zr | 1850 | 0.068 | 6.5 | very small | 91.22 |